Green Energy Alternatives

An Assessment of Free Market Energy Policy and the Role of the Federal

Government in the Promotion of Energy Efficiency and Commercialization of

Renewable Energy Technology

Dr Ted M Bixby

This book will analyze energy efficiency and renewable energy programs of the federal government. The analysis will focus on empirical data, expert testimony, past research and modeling to examine solar, wind, and marine energy and efficiency technologies. The environmental and economic impacts of renewable commercialization and widespread efficiency are reviewed. The research also addresses the role of the free market as it relates to federal energy policy and renewable commercialization efforts.

This book is dedicated to Elizabeth CJ and Brody.

Table of Contents

Chapter 1 Introduction to Oil Based Economics

Overview of Situation
Purpose of Research
Impetus for Research
Impacts
Summary
Definitions

Chapter 2 Analysis of the Oil Based Economy

Overview
Analysis of Approach
Transportation Sector
Construction Sector
Agricultural Energy
Economic Benefits of Sustainable Energy
Green Power Markets
The Electricity Grid
Environmental Impacts
Climate Change and Energy Policy
Wind Power
Sun Power
Hydropower
Micro Power
Sustainable Land Use
Biopower
Energy Efficiency
Dependence on Oil

Chapter 3 Methodology and Analysis of Green Energy Alternatives

Overview
Benefits of Methodology and Approach
Qualitative versus Quantative Analysis
Triangulation
Variables
Grounded Theory

Chapter 4 Data Assessment Alternatives to an Oil Based Economy

Overview
Solar Powers
Marine Energy
Wind Power
Summary

Chapter 5 Results and Final Thoughts

Summary
Significance
Limitations
Recommendations
Conclusions

Chapter One

Introduction to an Oil Based Economy

This chapter provides an overview of the problems surrounding present federal energy policy related to energy efficiency and renewable energy sources. The fundamental problem with the status quo is the lack of a long range coordinated energy policy sponsored by the Federal Government. The Federal Government's policies are piecemeal and ad hoc; they fail to provide a clear unified directive to the many participants in the energy sector. The government has failed to adequately address the over reliance on fossil fuels as a long term fuel source and present policies are woefully inadequate when it comes to promoting energy efficiency and the large scale commercialization of renewable energy technology.

Overview

The purpose of this research is to provide insight into the challenges in the United States energy sector, paying particular attention to energy efficiency and renewable energy technologies. This research and analysis investigation will highlight the challenges and opportunities encountered in introducing energy efficiency and renewable technology in to the marketplace. Present policy provides a financial disincentive to commercial renewable developers and excessive red tape bogs down many green energy projects. Through a qualitative assessment of the problem this book will coordinate and evaluate the existing literature and propose policy recommendations based upon the findings.

One of the greatest challenges facing our generation is to establish a reliable long term source of sustainable energy. The energy situation is inherently linked to the economic health of the nation in a complex myriad of ways. It is vital that the government make the correct decisions to ensure that energy sources remain economically and environmentally viable. The leadership position of the federal government makes policy action at this level indispensable to the long term success of any policy action. The present oil based economy is operating with a finite resource as the linchpin of our economic and energy policy. The government policymakers and business leaders must address the situation; and find acceptable energy alternatives long before the oil runs out in order to assure a smooth viable transition away from the fossil fuel economy. The scope of the study is relegated to an examination of the alternatives to an oil based economy.

Specifically, the viability of energy efficiency, conservation and renewable energy sources will be examined in light of recent technological and economic developments. The research will look at the practicality and feasibility of implementing broad scale efficiency technologies and the widespread commercialization of alternative/renewable fuel technologies. In particular the research will look at energy efficiency, renewable technologies as they relate to the major energy markets of the US economy. The rationale of the study and targeted research are straight forward: the paper seeks to assess the current state of energy policy and technology with the end goal of assessing effectiveness and making recommendations for future energy policy. The study will highlight areas for improvement and will suggest which technologies currently employed should be continued and emphasized moving forward.

Three major qualitative assumptions will be assessed during the course of this research. **First**, it will be argued that a new energy policy must include energy efficiency and renewable energy technologies as mainstays in order to assure the US has a reliable source of energy moving into future. **Second,** the theory will be researched and tested that a new non fossil fuel based economy will provide a number of economic benefits to the United States and more broadly the world economy. **Third,** it will be argued that a new energy policy with renewable power sources conservation and energy efficiency as cornerstones will alleviate many of the environmental problems associated with the current dependence on fossil fuels.

Purpose of Research

The research will be examined and discussed to determine the validity of the preceding theories. The analysis will focus mainly on empirical data and examination of existing energy technologies. First, it will be argued that a transition to energy efficiency and conservation is economically and technically feasible in today's oil based climate. The high price of oil and the extraordinary operating costs of fossil fuels based technologies combined with emerging economies of scale surrounding renewable energy sources have created a situation which is ripe for the exploration of alternative energy sources. A number of alternatives to fossil fuel will be examined and their viability will be assessed through empirical examples, computer modeling and extrapolation.

Second, the research will be analyzed to determine if a transition to a sustainable energy future will lead to positive economic and environmental changes for the United States and more broadly for the world as a whole. Specifically investment and support of commercialization of renewable and energy will lead to investment and employment opportunities for the US economy. Investment in these opportunities will lead to a lessening of trade deficits by decreasing dependence on foreign sources of energy. In fact dependence on foreign oil will lead to economic growth and a decrease in military spending. Lessening our dependence on foreign oil will decrease inflationary pressures within the economy and eliminate much of the demand pull inflation related to oil pricing.

Energy efficiency and renewable technology will potentially provide a number of benefits to the environment of the US. Moving towards a non fossil fuel based economy will allow the nation to avoid many environmental problems which are the result of the present fossil fuel based economy. The switch away from fossil fuels will lead to a lessening of pollution and environmental degradation. Specifically, the incidence of air pollution and acid rain will be decreased leading to lower health and environmental cost associated with these types of pollution. The rate of climate change will be slowed allowing the global economy and environment time to successfully manage the effects of these changes and navigate the risks with greater ease. This point should not be understated as the rate of change is crucial to managing any short or long term effects of global climate change. Indeed, if the rate of change can be managed through switching to sustainable energy sources it may be possible to successfully mange any adverse impacts which are the result of changes to the earth's atmosphere.

Renewable energy technology and widespread measures aimed at increasing energy efficiency are economically and environmentally attractive alternative to a fossil fuel based economy. Implementation of energy efficiency measures while transitioning to commercial renewable technology installation will provide the United States with the most economically effective energy policy when examined against the alternatives. The research will suggest that the time is now to perform such an assessment of present policies. To forge a new long term energy policy which is both successful and comprehensive the sooner the process is undertaken the better.

The challenge of providing energy may be one of the most fundamental issues faced as the economy heads into the 21st century. The U.S. economy depends on an adequate source of fuel supplies in order to maintain effective economic growth and stability. This research will look at the interrelationships between energy sources and free market economics. Particular attention will be paid to the relationship between energy efficiency technology, commercial renewable installations, and economic expansion. The research will highlight the potential benefits of large scale commercialization of renewable energy technology. The paper will examine the practicality of adopting these efficiency and renewable technologies in a broad and timely fashion. The research will examine what efforts are needed at the federal, state, and local levels. Suggestions will be made concerning necessary changes at all three of these levels.

The review is of vital importance given the relationship between reliable cost effective energy sources and the United States economy. The present oil based economy is limited by finite resources and eventually a transition away from fossil fuels must begin in earnest if the U.S. is to enjoy a smooth transition in to a free market which is based on renewable energy sources. Again this point should be emphasized; a transition away from oil is inevitable, there is no present technology which allows for the manufacture of fossil fuel through manmade processes. The geological fossil fuel based resources are finite and strategy changes must incorporate this notion if these strategies are to have any chance of success. The research will suggest and uncover a future which demands greater attention be given to energy efficiency in order to encourage conservation, economic stability, and a successful transition to a renewable energy based economic system.

The study will conduct an assessment of the present state of our fossil fuel economy and look at the short and long term implications for the environment and the economy. A number of factors are combined to create the critical mass necessary to allow the United States to transition in a new direction away from the present fossil fuel based economy.

Pioneering research and development by both the government and the private sector have yielded a host of promising new technologies that turn abundant domestic energy sources—including solar, wind, geothermal, hydro, biomass, and ocean energy—into transportation fuels, electricity, and heat. However, there is a leadership vacuum in the area of energy policy and some analysis is necessary in order to determine what kind of guidance is necessary and who should provide the necessary leadership. There are issue of logistics and funding which need to be assessed so that a successful groundwork for future energy policy can be established.

The jury is still out on which course of action will be the best however it is clear that the research needs to be consolidate and assessed so that the decision making process can be refined and enhanced. This paper and the analysis which drive it are critical in order to provide a clear picture of present energy policy and in order to provide some clarity to the decision making process. Many of the new technologies that harness renewable energy sources are economically competitive with the fossil fuels. In order to avoid repeating the mistakes of the past, a thorough analysis and a qualitative data assessment is required. Fossil fuels are still the main stay of the US economy and energy policy; however the time is rapidly approaching when an informed revision of energy policy will be required in order to lead our energy policy into a future based on renewable sources of energy.

Impetus for Research the Need for Green Energy

With oil prices soaring, the security risks of petroleum dependence growing, and the environmental costs of fossil fuels becoming more apparent, there are compelling reasons to put renewable technologies to use on a large scale. Energy policies must be scrutinized and assessed in order to determine their long term viability. Energy transitions take time, and no single technology will solve energy problems, therefore it is essential that a comprehensive examination of available alternatives is undertaken. It is quite possible that renewable energy technologies, combined with substantial improvements in energy efficiency, have the potential to gradually transform the U.S. free market away from dependence on fossil fuels.

A new and better energy future is possible if the country can forge a compelling vision of where it wants to be focused moving forward. Yet is obvious to the astute observer that a comprehensive assessment is required if the proper strategies are to be implemented in a timely fashion. Recent developments in the global marketplace show the potential of renewable energy for example, global wind energy generation has more than tripled since 2000, providing enough electricity to power the homes of about 30 million Americans. It is critical that we examine the localized success of these commercial installations. Understanding the driving forces behind these transitional shifts in the marketplace are crucial to developing policies in the future which adequately mirror the success of the past. The United States led the world in wind energy installations in 2005 and an open honest assessment is required to determine what the logical next steps should be in formulating a comprehensive energy policy.

.

Global investment in renewable energy (excluding large hydropower) in 2005 is estimated at $38 billion—equivalent to nearly 20 percent of total annual investment in the electric power sector. A more careful examination of this activity is essential if past success is to be replicated and the mistakes of the past are to be avoided. There is a groundswell of support for investments and projects to be established in commercial renewable technology and building energy efficiency. If large scale commercialization of these energy technologies is to occur a comprehensive analysis is required to assure the viability and successful implementation of this commercial industrialization. (Renewable Energy Policy Network 2008)

Renewable energy investments have nearly doubled over the past three years, and have increased six-fold since 1995. Next to the Internet, new energy technology has become one of the hottest investment fields for venture capitalists. Now is the time to provide a qualitive analysis which can serve as a guide for future policy formulation and decision making. These dynamic growth rates in the field of renewable technologies are driving down costs and spurring rapid advances in technologies. They are also creating new economic opportunities for people around the globe. With adequate research and understanding the success of future efforts can more adequately be ensured. Today, renewable energy manufacturing, operations, and maintenance provide approximately two million jobs worldwide. In order to provide replication of this success and future opportunities in the marketplace research is required to assure the government embarks on the right path in designing energy policy. (REPP Archives 2008)

In order to break the national addiction to outdated fuels and technologies, the Federal Government will need a renewable based energy efficient energy policy energy policy. The prominent positions that Germany and Spain hold in wind power, and Japan and Germany enjoy in solar energy, were achieved thanks to strong and enduring policies that their legislatures adopted in the 1990s. These policies created steadily growing markets for renewable energy technologies, fueling the development of robust new manufacturing industries. By contrast, U.S. renewable energy policies over the past two decades have been an ever-changing patchwork. Abrupt changes in direction at both the State and Federal levels have deterred investors and led dozens of companies into bankruptcy. To join the world leaders and achieve the US free market's full potential for renewable energy, world-class energy policies based on a sustained and consistent policy framework at the local, state, and national levels will be required.

Expensive oil is spurring rapid advances in technologies. Renewable energy manufacturing, operations, and maintenance provide approximately two million jobs worldwide. The United States will need a much stronger commitment to renewable energy in order to take advantage of future opportunities. Dependence on fossil fuels is rising, even in the face of high oil prices and growing concern about global warming. Of particular concern is the well over 100 coal-fired power plants now on the drawing boards of the U.S. electricity industry—most of which lack the latest pollution controls and could still be pumping carbon dioxide into the atmosphere a half century from now. (Environmental and Energy Study Institute 2008)

Several states are demonstrating just how quickly renewable energy can take hold with the right policies. California already presently gets 31 percent of its electricity from renewable resources. Texas, whose history is closely identified with the oil industry, now has the country's largest collection of wind generators. Iowa produces enough ethanol that if this were all consumed in-state, it would meet half the state's gasoline requirements. A national coalition of more than 200 business and citizens organizations—led by the farm and forestry sectors—has proposed a national commitment to obtaining 25 percent of U.S. energy from renewable resources by 2025. A new economic analysis by the Rand Corporation for the Energy Future Coalition concludes that if the United States were to get 25 percent of its electric power and transportation fuels from renewable energy by 2025, the country's energy costs would be reduced, with large savings occurring by 2015. And national carbon dioxide emissions would fall by one billion tons. (Energy Future Coalition 2008)

Impacts of Change in Policy--Benefits of Green Economy

A number of changes would be readily observable if the United States federal government were to promote a shift towards energy conservation and renewable energy technology. In the event that the United States implement a comprehensive policy aimed at promoting energy efficiency and supporting the commercialization of renewable energy technology the following results would be expected.

The energy economy would become more decentralized and efficient, allowing homes and businesses to meet many of their own energy needs. Dependence on imported oil would decline, improving U.S national security. Trade deficits would fall as oil imports decline, reducing the roughly $300 billion the United States spends on foreign sources of oil. The air would be cleaner, reducing asthma and other respiratory diseases and saving American lives. Emissions of global warming gases would decline, reducing the threat to cities and coastal properties from rising sea level and the threat to agriculture from drought and higher temperatures. Hundreds of thousands of new jobs would be created in the agricultural, manufacturing, and service companies that would emerge to meet the demand for renewable energy. Rural communities would be revitalized as farmers and ranchers, who own the land where much of the renewable energy can be harnessed, would reap the benefits.

Present energy policy fails to fully utilize and appreciate the potential of energy efficiency renewable and conservation technologies. Natural resources such as coal, oil, or natural gas, take millions of years to form naturally and cannot be replaced as fast as they are consumed. At present, the main energy sources used are non-renewable; renewable resources, such as solar, tidal and geothermal are yet to be fully utilized. The scarcity of fossil fuel based resources has been repeatedly documented by University scholars, policy makers and energy officiandos. The world is faced with a limited finite amount of fossil fuels and must face the reality that one day these resources will no longer exist as they do in their present abundant state. The future must contain an energy policy based on efficient energy and the commercialization of renewable technology.

The free market economy has always prospered by facing daunting challenges and transforming them into opportunities for innovation, industry, and growth. The federal government has the ingenuity, know-how, and determination necessary to create an energy-secure America. With research behind clear policy guidelines and coordinating efforts between the public and private sectors, the government can lead a coalition to find exciting new ways to build free market use of domestic, non-polluting renewable energy. By capturing the energy of the wind and the light of sun, the power of a mighty river or heat stored in the crust of the Earth, we can find new untapped resources that create jobs, improve our security, and build the health of our people, our planet, and our economy.

The research guiding this analysis will illustrate that an energy future based on abundant and clean renewable resources is not only urgently needed, but achievable. The time is ripe for a strong national commitment to enacting new policies at the Federal, State, and local levels that will allow the United States to become a world leader in building a 21st century energy system. The hypothesis and research proposals guiding this analysis will illuminate a path for mobilizing local communities, and government leaders to share in this vision for a clean, secure, and prosperous future with commercial renewable technology and increased energy efficiency.

The main thrusts of the research will argue that energy efficiency in the near term and renewable sources of energy in the longer term will prove to be the answer to the rapidly approaching question of what does life after fossil fuels look like. Energy efficiency improvements based on off the shelf technology can greatly assist the United States in meeting its energy needs immediately.

The problem in energy policy is the lack of a broad based policy which addresses the environmental and economic factors incorporated with fossil fuel dependence. Despite some decentralized local efforts there is still no comprehensive long term energy policy. While the technologies exist, the corporate and political will to put them into widespread use does not. Many companies in the automobile and energy industries put pressure on the White House and Congress to halt or delay new laws or regulations -- or even to stop enforcing existing rules -- that would drive such changes. From requiring catalytic converters to improving gas mileage, car companies have fought even the smallest measure to protect public health and the environment.

Summary

As the largest energy consumer in the United States, the federal government has both a tremendous opportunity and a clear responsibility to lead by example with smart energy management. By promoting energy efficiency and the use of renewable energy resources at federal sites, the Federal government can save energy, save taxpayer dollars, and demonstrate leadership with responsible, cleaner energy choices. (Energy Efficiency and Renewable Energy 2008)

The Federal Government and the free market face grave challenges in the field of energy—from the gathering storm of global warming to a dangerous addiction to oil that jeopardizes our national and economic security. By promoting renewable energy and capturing the energy of the wind and the light of sun, the market can find new untapped resources that create jobs and build the health of the economy.

Green energy alternatives will illuminate a future economy based on abundant and clean renewable resources are not only urgently needed, but achievable. The time is ripe for a strong national commitment to enacting new policies at the federal, state, and local levels that will allow the United States to become a world leader in building a 21st century energy system. Meeting that challenge will require concerted action by governments, businesses, and citizens across all sectors of the energy market. All but four U.S. states now have incentives in place to promote renewable energy. More than a dozen have enacted new renewable energy laws signaling fresh political momentum. If such policies continue to proliferate, and are joined by Federal leadership, including financial incentives promoting renewable technology and energy efficiency, rapid progress is possible.

Chapter Two

Literature Review

This chapter will provide an insight into existing body of literature as it relates to energy efficiency and renewable energy technology. The review will help establish the grounds to proceed forward with a qualitative exploration of alternative energy policies. An overview of major reports and current schools of thought regarding these technologies will allow us to gain a greater understanding into the practical applications of these energy technologies. The review will look at expert testimony, empirical data, and scholarly research and computer models.

Overview

The current literature offers a mixed bag of results when examining the economic validity of energy efficiency and renewable energy. A number of major studies are reviewed to determine the overall appeal and attractiveness of existing technologies and energy policy. There are a number of areas which are pertinent and need further research such as free market acceptance and promotion of renewable technologies. Many studies today are narrowly focused on a singular technology and conversely many studies are too encompassing in their scope causing some issues to receive only terse coverage. First, this research will focus on energy efficiency as it relates to the economy and investment markets. Second, the viability of energy efficiency and renewables as a way to transition away from a fossil fuel economy will be examined. Analysis based a number of different methods will be reviewed to evaluate the effect of energy efficiency technologies and renewables would have as an alternative to the present fossil fuel based economy. Third, the effect of energy efficiency and renewable related to their role in reducing pollution will be evaluated.

Analysis of Approach

The methodology of the research is based on some pretty straightforward assumptions and principles. Each energy technology was examined using multiple criteria to determine effectiveness and appeal of the technology. The research examines the current market penetration of the technology and the underlying cost drivers related to commercialization and implementation. The examination also looks at the opportunity technologies provide in terms of broad scale commercialization and application of these energy efficiency and renewable energy technologies. The research further estimates the potential for technology improvement by triangulating information from a number of sources including computer modeling, expert testimony, empirical example and technology assessment. Finally, the research will examine the potential for implementation of these technologies and examines their viability when contrasted against existing technologies.

The literature review is focused in a few core areas as it relates to energy policy. The analysis will examine the potential for energy efficiency technologies in three main areas: construction, transportation, and agriculture. The research regarding renewable energy technology concentrates on solar wind and tidal energy. The analysis will deliver a reasoned assessment of the potential for energy efficiency and renewable technology to serve as both a short term and long-term replacement for our present utilization of fossil fuel based energy sources. The review will examine the practical economic and societal effects of transitioning away from finite fossil fuel technologies toward a future which includes sustainable renewable technology. The research will examine our present energy policy and what are the drivers behind this policy. The research will then move into an assessment of the present state of energy efficiency technology and look at the drivers and keys to success and/or failure of energy efficacy technology may be. The last part of the literature review will look at renewable energy technology and its viability in today's marketplace. The review will bring to light some of the major advantages of renewable sources of technology when examined vis a vis fossil fuels technology.

The research presented here differs from previous reviews in two critical areas. First, the scope of this research is limited to energy efficiency technology in three main areas/ sectors transportation, construction, and agriculture. The review also limits the scope of renewable technology to solar wind and tidal power. Secondly, the research provides an in-depth analysis of the economic effects of the technologies. There has yet to be a comprehensive examination of renewables and energy efficiency technologies which is solely focused on highlighting the economics surrounding these technological applications. Many studies have examined pollution related cost and have looked at the environmental effects/benefits of these technologies but no study has yet to uncover fully the economic effects of adopting energy efficiency technologies or the effect of broad scale commercialization of renewable technologies.

Transportation

Encouraging greater energy efficiency in the automobile and commercial transportation market sectors is critical if we are to make headway in the battle to reduce dependence on fossil fuels. The transportation sector is one of the most important markets because of its direct impact on the demand for oil. Transportation accounts for two-thirds of U.S. oil consumption and is the predominant source of domestic urban air pollution. Recent gasoline price increases have combined with growing environmental concerns to spur interest in new fuels to run the nation's transportation fleet. The present market forces and government policy are not addressing the situation. The current transportation mix across sectors relies on oil for more than 95 percent of its energy. Renewable fuels currently represent only around 2 percent of the total. The immediate options for running the U.S. transportation system on renewable energy are more limited than those for other sectors of the economy, such as buildings and industry. However, this is an important sector for action because it is clearly observable to the public. (American Council for Energy Efficient Economy 2008)

In the short term, the main potential is in the use of biofuels derived from crops and wastes. In the long term, electricity and hydrogen derived from sources like wind and solar energy are likely to become viable alternatives. A federal law provides tax credits for purchasers of hybrid and alternative fuel vehicles. Many states also offer incentives for buying these vehicles. The same "green" consumers who have made hybrid gas-electric vehicles hot items in auto showrooms in recent years are now showing strong interest in biodiesel and other renewable fuels.

The market seems to be steadily evolving in favor of more fuel efficient vehicles however the success has been on a limited scale. Electric cars on the market today can be plugged into an outlet and recharged at home. Homeowners with rooftop solar systems—or in regions rich in hydro or wind power—can already fuel their vehicles with renewably generated electricity. Given existing technology and projections available from experts it appears that a new generation of plug-in hybrids will soon provide a similar opportunity, while giving drivers the option of extending the typical 100-mile range of an electric vehicle by using gasoline or biofuels in the tank. (American Council on Renewable Energy 2008)

In the future, hydrogen offers a means of storing energy sources such as solar and wind power. Hydrogen can be produced from water using any energy source that generates electricity. Because it can be readily stored in tanks and transported in a pipeline, hydrogen is a logical long-term replacement for oil and natural gas. A new generation of experimental fuel-cell vehicles is being developed that efficiently uses hydrogen to turn the wheels, with water vapor the only tailpipe emission. As renewable energy becomes a larger part of the electricity system and as costs decline, renewably generated hydrogen is likely to become a growing part of the transportation fuel mix. It is important for the government to pave the way for hydrogen's success by subsidizing existing free market efforts and sponsoring joint research with private sectors hydrogen entrepreneurs.

Construction Sector

The construction sector allows enormous possibility for energy savings. The opportunity is available in the short term, as many of the technologies required to make these efforts successful can be achieved with off the shelf technology. Commercial and residential buildings consume about one-third of all U.S. energy and two-thirds of U.S. electricity. In addition, the commercial and residential construction sectors account for more carbon emissions than any other sector. After thorough review it is obvious that most buildings' demand for energy can be dramatically reduced, and renewable energy can meet a significant share of the remaining energy needs.

The grass roots green building movement seeks to tap consumer demand for environmentally friendly, healthy, and affordable homes and offices. Designers of green buildings aim to minimize energy consumption with more-efficient materials and appliances and integrated renewable energy systems. Designers seek to reduce demand for water and open space, to use sustainably produced products (including recycled materials), and to provide convenient access to public transportation. Green Builders propose low impact construction which employs both energy efficiency and renewable technologies as centerpieces of the construction plans. This is the type of movement which should be fostered and assisted by government. This Non Governmental entity can be assisted by funding and easing of regulations which will allow this type of construction to go on unabated without bureaucracy getting in the way. It will take both government support and market demands in order to further propel the green building movement.

The green building movement officially began with the founding of the U.S. Green Building Council. In 2000 the Council published LEED (Leadership in Energy and Environmental Design) standards to guide developers' decisions on site design, water use, indoor air quality, and energy generation and use. Nearly 6,000 member organizations and companies plan to construct new buildings or renovate old ones according to LEED standards, and a growing number of state and local governments— including in Atlanta, Boston, and San Francisco—have incorporated them into laws and regulations for new public buildings. (Green Building Alliance 2008)

Solar energy is playing a role in the construction and success of many of these buildings. The pharmacy chain Walgreens plans to install solar photovoltaic technology on 112 of its stores. This will enable Walgreens facilities to meet 20–50 percent of their power needs on site. Another successful example of green construction can be found in Battery Park in New York City. In this location developers built the world's first green high-rise. The "Solaire" apartments use 35 percent less energy and 65 percent less electricity than an average building, with solar cells meeting at least 5 percent of demand. By 2009, all developments covering Battery Park City's 92 acres will be LEED certified and will have solar panels. (Green Building Alliance 2008)

The Chicago Center for Green Technology uses geothermal energy for heating and cooling, and the Dallas/Fort Worth Airport relies on solar energy for air conditioning, reducing cooling costs by 91 percent at times of peak demand. Major housing developers such as Centex and Premier Homes are now incorporating solar into new homes in California. There are good economic reasons for constructing green buildings, which generally have healthier employees, higher worker productivity, lower turnover, and significant energy and water savings. More Examples of Green Buildings in the United States Ford Motor Company installed a green roof on the 10.4- acre rooftop of its Rouge River Plant in Michigan in 2004. Replacing dark, heat-absorbing roof surfaces with plants keeps buildings cooler in summer and warmer in winter, reducing energy use for heating and cooling by 10–50 percent; it also filters the air and rainwater. Pittsburgh's David L. Lawrence Convention Center includes numerous features that reduce the energy bill by at least one-third, or enough to meet the needs of 1,900 households. Its curved roof allows hot air to escape through vents and cool breezes to flow in from the river. Construction costs were comparable to or lower than other (non-green) centers built in recent years. Genzyme's headquarters in Cambridge, Massachusetts, was the first large U.S. office building to achieve "platinum" LEED standards, the highest level of certification. The building includes a green roof, uses natural light and ventilation, is sited on a reclaimed brown field and close to a subway station, and provides indoor bike storage, showers, and lockers for employees. (Green Renewable Electricity Certification Program 2008)

A study by the California Sustainable Building Task Force found that an upfront investment of 2 percent (the average cost premium) in green building design results in average savings of at least 10 times the initial investment over a 20-year period. Costs are falling as those who design, construct, and maintain green buildings move further along the learning curve. Further, green buildings tend to have higher occupancy rates and rents, and therefore better returns on investment, than conventional buildings. And generating power and heat on-site with renewable energy can reduce the chances of a power outage, while hedging against an increase in electricity prices. (Green e Renewable Electricity Certification Program 2008)

Energy for Agriculture

Renewable energy opportunities in the farming industry are numerous and provide a critical area for the government to realize energy savings and begin a transition away from fossil fuels. Renewable energy could provide a new source of revenue for thousands of farmers and agricultural processors, creating economic opportunities in rural areas that have suffered from decades of falling crop prices. Many farmers are benefiting from the recent high demand for ethanol and research shows that the growing ethanol and biodiesel industries are providing jobs in plant construction, operations, and maintenance. According to the Renewable Fuels Association, the ethanol industry created almost 154,000 U.S. jobs in 2005 alone. The result of this job creation was enough to boost household income by $5.7 billion. It also contributed about $3.5 billion in tax revenues at the local, state, and federal levels. (Renewable Fuels Association 2008)

Farmers and rural communities can also increase their revenue by tapping local wind resources to generate electricity. Opportunities in wind power technology abound for those in prime installation locations. Some of the country's most valuable winds sweep across some of its poorest farmlands. Here wind power technology may represent one of the few options available to these agricultural economies. Installation of wind farms will allow these farmers and ranchers to generate income even when cropland is parched from drought. Farmers can become wind developers themselves, or opt to have commercial developers install turbines on their land and, in turn, receive annual lease payments or share the revenues from a wind project. The payback from a wind farm is twofold. First, it is an energy provider and a steady source of revenue for the ailing agricultural sector. Wind power payments range from $1,000 to $4,000 a year for each wind turbine installed, as much as doubling the economic yield from the land. Second the wind farm installation does not prevent the farmer from the normal scope of operations, farmers and ranchers can continue to raise crops and livestock beneath them. (American Wind Energy Association 2008)

Solar energy benefits farmers in many ways, by lighting and heating buildings and greenhouses, drying crops, and powering water pumps and irrigation systems. Empirical examples of successful efforts can be found in California where one of the state's largest vegetable growers now irrigates 600 acres of farmland with solar power, helping to ease pressure on the California electricity grid during peak demand periods. The time is critical now for government policies to swing in line with public opinion, so that construction, agriculture, and transportation are able to operate efficiently and sustainably. (American Solar Energy Society 2008)

Economic Benefits

Creating employment opportunity through implementation of sustainable energy policy is one of the prime economic benefits resulting from the implementation of energy efficiency technology and commercialization of renewable technologies. Expanding the use of renewable energy will have a positive impact on employment, according to more than a dozen independent studies analyzing the impact of clean energy on the economy. Renewable energy creates more jobs per unit of energy produced and per dollar spent than fossil fuel technologies do. When accounting for all variables, several studies have shown that greater reliance on renewable energy would have large, positive impacts on the U.S. economy, creating significant numbers of new jobs, driving major capital investment, stabilizing energy prices, and reducing consumer costs. (RFA 2008)

A transition away from fossil fuels and toward renewable energy would create both winners and losers, but on balance most studies show that many more jobs would be created than lost. A 2004 analysis by the Union of Concerned Scientists found that increasing the share of renewable energy in the U.S. electricity system to 20 percent—adding more than 160,000 megawatts (MW) of new renewable energy facilities by 2020—would create more than 355,000 new U.S. jobs. This is exactly the type of economic opportunity many local economies need if they are they are to remain economically relevant and viable. (Union of Concerned Scientists 2008)

A potential snowball effect will occur if the increased use of renewable energy led to significant reductions in fossil fuel prices, consumer savings on electricity and natural gas bills would ripple through the U.S. economy, spawning even more jobs. The savings would lead to the creation of additional discretionary income which would translate in to immediate employment gains. This economic activity would also potentially represents a tremendous economic boost to rural communities. Most of the jobs created in renewable energy would be high-paying positions for skilled workers, in fields such as manufacturing, sales, construction, installation, and maintenance. This type of employment opportunity is critical as many local economies look for ways to replace lost jobs in the manufacturing sector. (Rocky Mountain Institute 2008)

A 2004 Renewable Energy Policy Project study determined that increasing U.S. wind capacity to 50,000 MW—about five times today's capacity level—would create 150,000 manufacturing jobs, while pumping $20 billion in investment into the national economy. Renewable heating and biofuels also offer significant employment opportunities. The U.S. ethanol industry created nearly 154,000 jobs throughout the nation's economy in 2005 alone, boosting household income by $5.7 billion. The opportunity is sizeable and the success of the past is certainly something that bears repeating. (Renewable Energy Policy Project 2008)

By contrast, the economic outlook for the fossil fuel industry is bleak. Employment in the fossil fuel industries has been in steady decline for decades, in large measure due to growing automation of coal mining and other processes. Between 1980 and 1999, while U.S. coal production increased 32 percent, related employment declined 66 percent, from 242,000 to 83,000 workers. The coal industry is expected to lose an additional 30,000-some jobs by 2020, even if coal demand continues to rise. Further, high prices for fossil fuels have a negative impact on the economy, even leading to the transfer of manufacturing jobs overseas. Expanding the use of renewable energy can help minimize these losses and provide new opportunities for displaced workers. The implementation of energy efficiency technology will also assist in generating additional employment opportunities. (Center for Resource Solutions 2008)

Most of the investment to date has occurred in a relatively small number of countries, driven by consistent, forward-looking policies that aim to create markets for renewable energy. Germany and Spain, for example, have forged a dominant position in wind energy over the past decade, and are now turning to other renewables as well. Japan and Germany lead in solar electricity, with Japan responsible for nearly half of global solar cell production and Germany dominating the marketplace. Brazil has moved to the forefront of biofuel production with its successful alcohol fuels program. And China is the world leader in small hydropower and solar water heating, with well over half the global market in each. Despite strong public support and rapidly rising interest in renewable energy, the United States has not kept up with the strong growth in renewables over the past decade; as a result, its market share has fallen steadily. For example, while U.S. solar cell manufacturing has risen year by year, the nation's share of global production has declined from 44 percent in 1996 to below 9 percent in 2005. (Worldwatch Institute 2008)

Time is growing short for the United States to get back in the game and compete for what could be some of the largest new markets of the next few decades. A strong partnership between government and the private sector is essential if that kind of leadership is to be achieved. Taking steps now is critical to building the foundations of a sustainable energy policy. By adopting these technologies now the maximum return will be realized and the full economic potential of the technology will be felt throughout the US economy. (Renewable Energy Policy Project 2008)

Green Power Markets

The effect of sustainable energy policy on the provision of electrical and economic markets has begun to make an impact. The effect of green sustainable power is making itself felt throughout the US energy grid. Voluntary purchases have played a major role in driving the U.S. renewable energy market. By the end of 2004, green power demand had topped 2,200 MW of renewable capacity, up from 167 MW in 2000.

A number of examples demonstrate the empirical success of renewable energy based green markets. The U.S. Air Force is the nation's leader in green power purchasing, followed by Whole Foods Market and a growing list of corporate and government offices. The Statue of Liberty in New York City now gets 100 percent of her power from renewable energy. In most cases, green power subscribers pay a premium price for electricity, but some customers in Colorado and Texas are now paying less than non-subscribers due to rising natural gas prices. (Clean Energy Group 2008)

Investment opportunities in the broad scale commercialization of efficiency and renewable technologies are significant. Annual global investment in "new" renewable energy has risen almost six-fold since 1995, with cumulative investment over this period of nearly $180 billion. The $38 billion invested in renewable in 2005 compares to the roughly $150 billion invested worldwide in the conventional power sector in 2004. Market growth has been driven by technology improvements, rising fossil fuel prices, government policies, and the growing familiarity of investors and lenders with the opportunities and risks posed by the wide range of renewable technologies and projects. Renewable energy technologies tend to be more capital intensive than traditional fossil fuel technologies, with higher upfront costs. At the same time, renewable energy projects do not expose owners to the risks of fuel price increases or the cost of future retrofits or penalties associated with environmental and health problems. As a result, renewable and fossil fuel projects have very different financial profiles. (Renewable Energy Policy Project 2008)

Investing in renewables is no longer just about doing the right thing; it's also about making money. Renewable energy is increasingly viewed as an attractive investment by private and public equity investors alike. In November 2005, Goldman Sachs committed to investing more than $1 billion in renewable energy projects, including biofuels, solar power, and wind energy. The NASDAQ stock market launched its "Clean Edge U.S. Index" in May 2006 to track the performance of clean energy companies. In the world of venture capital, clean energy is the hottest new investment arena, having just passed semiconductors in annual deal flow, according to the Cleantech Venture Network. (Center for American Progress 2008)

Project lenders, principally banks, are providing loans to ethanol plants, wind farms, and other large-scale renewable power projects, and direct lending by U.S. banks and institutional investors is on the upswing. Still, U.S. banks lag behind those in Europe. One reason is that the financing of renewable energy projects in the United States is dominated by equity investments by the unregulated subsidiaries of electric utility companies, which benefit from the Production Tax Credit (PTC). The PTC has been available for wind power and certain waste projects, and was expanded in late 2004 to include solar, biomass, and geothermal power plants. The scores of ethanol plants now under construction are being financed by a wide array of agricultural coops, corporations such as Archer Daniels Midland, and equity investors ranging from large institutions to Microsoft Chairman Bill Gates. (Renewable Fuels Association 2008)

Public sector financing of renewable energy projects has been evolving for several years and is likely to increase substantially in the near term. By mid-2005, 17 Clean Energy Funds worth nearly $3.5 billion had been established in 13 states to support renewable energy development through grants, subsidies, loans, and investments that often leverage private sector financing. Cities are getting involved as well, using bond financing for renewable energy and energy efficiency projects.

The Electricity Grid

Examining alternate energy sources in their quest to effectively provide a stable source of electrical power is a prerequisite to an effective energy policy. The importance of a reliable environmentally friendly power grid cannot be overstated. The U.S. economy, as well as public health and safety, depends on a reliable power system that provides electricity 24 hours a day, 365 days a year. For example, the costly disruptions resulting from the Northeast blackout of August 2003 were a powerful reminder of how dependent the country is on the reliability of large power plants and the transmission networks that connect them. Currently, the U.S. electric power industry now relies on large, central power stations, including coal, natural gas, nuclear, and hydropower plants that as an aggregate generate more than 95 percent of the nation's electricity. Over the next few decades, renewable energy could help to diversify the nation's bulk power supply. Already, renewable resources (excluding large hydropower) produce 12 percent of northern California's electricity. (National Hydropower Association 2008)

Most electric utilities operate a combination of baseload plants (usually coal and nuclear) that operate most of the time and others (mostly natural gas) that are utilized only when demand is high. Some renewable power plants can provide steady power on demand whenever it's needed— using geothermal, concentrating solar (with storage), and bioenergy. Other renewable power sources are intermittent, meaning they are available only when the sun is shining or the wind is blowing. However, even intermittent sources can add significant value to the system by providing electricity when it is most needed and most costly to produce with conventional sources. In many parts of the country, for example, periods of peak sunlight coincide with peak power demand for air conditioning. By using renewable sources to provide power during periods of peak demand the economic return of renewables and energy efficiency technologies is maximized.

All power systems rely on backup generators, since even baseload plants must close occasionally due to technical problems. This provides a perfect opportunity to utilize intermittent renewable power sources. In the case of intermittent renewables, wind resources can already be forecast at least two days in advance, and fluctuations in power output can be reduced if not eliminated by spreading solar or wind generators across a sufficiently wide region. Studies show that even when wind power alone provides 20 percent of the total electricity on a regional grid—as it does in Denmark and large parts of Germany—backup capacity is rarely needed. Above that level, some backup capacity may be required, but at much less than a 1:1 ratio. In the future, new technologies like advanced gas turbines and fuel cells, as well as new storage devices, will likely reduce the cost of providing backup capacity, allowing much higher levels of dependence on intermittent generators. (American Council on Renewable Energy 2008)

Another empirically observable benefit of these technologies is that they also provide grid operators with real economic benefits (in addition to their peaking value) that are just beginning to be recognized. Conventional power plants based on coal and nuclear power can take 5–15 years to plan and construct, a serious disadvantage given the uncertainties of future power demand and the risks of borrowing hundreds of millions of dollars while the plants are built. Construction lead times for large renewable projects are often in the range of 2–5 years, reducing the risk to utilities and allowing capacity to be added incrementally to match load growth. According to FPL Energy, it can take as little as 3–6 months from ground breaking to commercial operation with new wind farms. As an added benefit once on line, renewable facilities can begin operation more rapidly than conventional power plants after blackouts, reducing associated economic and security costs. (American Council on Renewable Energy 2008)

At a time when the price of natural gas, the most popular fuel for recently constructed power plants, has increased significantly and other fossil fuels carry heavy environmental and economic costs, renewable power has become a valuable component of a utility power portfolio. Renewable energy provides a reliable source of emissions free energy and a hedge against future fuel-price increases. Wind farms are already price competitive with gas and coal, and GE Wind has predicted that wind turbine sales could surpass gas turbine sales within the next few years. As a final benefit, since renewable power plants are emissions free, commercial renewable technology also represents a hedge against future environmental regulations, including possible caps on mercury and carbon dioxide emissions.

Environmental Impacts

Moving toward energy sources which inhibit environmental degradation will allow the US to move away from an economy which is powered by fossil fuels. The transition away from pollution based sources of energy towards emission free power can begin immediately as existing technologies are now available to be implemented. A number of options are now available as off the shelf technology which will allow renewable technology to have an immediate dramatic impact on the amount of pollution emitted from energy sources. Adopting energy efficiency technology in the short term and moving to renewable commercialization in the long term will allow an energy policy that is much healthier and environmentally friendly that the current fossil fuel based approach.

The emissions-free nature of most renewable energy technologies is one of their principle advantages compared to fossil fuels. Power plants, motor vehicles, and industries that burn fossil fuels emit a host of pollutants that imperil human health, impose heavy economic costs, and degrade the natural environment. A 2002 study published in the Journal of the American Medical *Association* determined that exposure to air pollution poses the same risks of dying from lung cancer and heart disease as does living with a smoker. A 2004 study by ABT Associates estimated that fine particulate pollution from power plants causes nearly 24,000 premature deaths annually in the United States. Thousands more Americans experience asthma attacks, and millions of workdays are lost annually due to pollution-induced illnesses. The result is more than $160 billion per year in medical expenses due to air pollution from power plants alone. (Clear the Air 2008)

Sulfur emissions, resulting primarily from the burning of coal in conventional power plants to produce electricity, are the main source of acid rain, which damages crops, forests, and buildings and can make lakes and rivers too acidic to support life. Nitrogen oxides (NOx) combine with other chemicals to form ground-level ozone, or smog. The burning of fossil fuels also releases volatile organic compounds. Some of these components combine with NOx to create smog; others are directly toxic and are associated with cancer, developmental disorders, and adverse neurological and reproductive impacts. Coal and oil contain toxic metals such as mercury, arsenic, and lead that are released into the air when these fuels are burned and find their way into drinking-water supplies. Coal-fired power plants are the nation's largest human-caused point source of mercury pollution, emitting about 48 tons into the air each year. They alone are responsible for 42 percent of the nation's mercury emissions. (Clear the Air 2008)

In August 2004, the head of the EPA warned that fish in nearly all of the nation's lakes and streams are contaminated with mercury. Studies show that one in six American women of childbearing age may have blood mercury concentrations high enough to cause damage to a developing fetus. Mercury damage can affect the central nervous system and may damage reproductive, immune, and cardiovascular systems. Conventional power plants require significant amounts of water for ongoing maintenance and cooling. Withdrawal of surface water can kill fish, larvae, and other organisms trapped against intake structures, while wastewater discharge releases chemicals and heat into surrounding ecosystems, affecting plants, fish, and animals. (Blatt 2005)

Fuel extraction and transport pose severe health and environmental threats as well. Black-lung disease kills an estimated 1,500 former coal miners annually. In the Appalachian states of

West Virginia, Kentucky, and Tennessee, mountaintop coal mining (which involves blasting away mountain tops to expose coal seams within) has buried or polluted more than 1,200 miles of streams, destroyed more than 7 percent of Appalachia's forests, and eliminated entire communities. If current trends continue over the next decade, affected land in this region will cover 2,200 square miles, an area larger than the state of Rhode Island. The European Union has found that environmental and health costs associated with conventional energy and not incorporated into energy prices equal an estimated 1–2 percent of EU gross domestic product, excluding costs associated with climate change. A dramatic increase in our use of renewable energy could significantly reduce these burdens and breathe life in to local economies. . (European Renewable Energy Council 2008)

The growing threat of air pollution is a menace facing many citizens across the United States. Air quality indexes across the country are at all time lows as air contaminants continue to accumulate. More than 150 million Americans—more than half the nation's people—live in areas where air quality threatens their health. The pollution takes an enormous toll o the U.S. economy. A 2005 study by the Mount Sinai School of Medicine's Center for Children's Health and the Environment estimated that the cost in lost productivity to the U.S. economy due to mercury's impact on children's brain development totaled $8.7 billion per year. (Clean Energy States Alliance 2008)

Climate Change and Energy Policy

Most renewable energy sources add little or no carbon dioxide (CO_2) to the atmosphere. These renewable and energy efficiency technologies are therefore one of the key elements of a global strategy to reduce the threat of climate change. Atmospheric CO_2 concentrations have climbed 20 percent since measurements began in 1959 and nearly 36 percent since the dawn of the Industrial Revolution. Over the past century, the average global temperature has risen by 1.8 degrees Fahrenheit; more than half of this warming has taken place in the past 30 years. The burning of fossil fuels for energy production and use is responsible for an estimated 70 percent of the global warming problem, and the United States accounts for about one-quarter of total global emissions. (Climate Solutions 2008)

In its 2001 report, the Intergovernmental Panel on Climate Change, the most authoritative scientific body synthesizing the vast research on climate change, concluded that "there is new and stronger evidence that most of the warming observed over the last 50 years is attributable to human activities." Expected impacts of global warming include sea-level rise; flooding of coastal areas; increased frequency and severity of floods, droughts, storms, and heat waves; reduced agricultural production; massive species extinction; and the spread of vector-borne diseases such as malaria and dengue fever. (Appenzeller 2004)

The rate of climate change is alarming and there is growing concern that societies and ecosystems will not have time to adapt to these changing conditions. Rising economic losses due to weather-related disasters are part of a trend being linked to climate change. The World Health Organization estimates that climate change is already responsible for 150,000 deaths annually.

While developing countries will likely see the highest toll, impacts will be significant in industrial nations as well, including the United States. (Nicklen 2007)

The concentration of CO2 in Earth's atmosphere is now higher than at any time in the past 650,000 years, and the rate of increase is accelerating. In June 2004, a new, more-accurate atmospheric model revealed that global temperatures could rise more rapidly than previously projected. The extent of warming by the end of this century will be determined by the amount of fossil fuels we continue to burn and the sensitivity of the climate system.

U.K. Chief Scientific Advisor David King has said that climate change is "the most severe problem that we are facing today—more serious even than the threat of terrorism." At their July 2005 meeting in Gleneagles, Scotland, G-8 leaders issued a statement acknowledging that "climate change is a serious and long-term challenge that has the potential to affect every part of the globe." Former U.S. president Bill Clinton has warned that climate change "has the power to end the march of civilization as we know it," adding that a "serious global effort" to promote clean energy is required. (Appenzeller 2007)

Global emissions of greenhouse gases must be reduced dramatically over this century to avoid catastrophic climate changes. The sooner societies begin to reduce their emissions; the lower will be the impacts and associated costs of both climate change and emissions reductions. The Kyoto Protocol, which entered into force in early 2005, requires 39 industrial nations to reduce their emissions. Although the United States is not party to the treaty, U.S. companies that operate within signatory countries face pressure to reduce their emissions as well. Dramatically increasing the use of renewable energy, alongside significant improvements in energy efficiency, will provide an important means of doing so. (Clean Energy Group 2008)

Wind Power

The wind is one of the country's most abundant energy resources. About one-fourth of the total land area of the United States has wind activity powerful enough to generate electricity that is price competitive with natural gas or coal at today's prices. According to government-sponsored studies, the wind resources of Kansas, North Dakota, and Texas alone are sufficient to provide all the electricity the nation currently uses. Although wind power presently provides less than 1 percent of U.S. electricity, it is poised to expand dramatically. (Bolinger 2007)

Wind energy technology has advanced steadily over the past two decades. Average turbine size has increased from less than 100 kW to more than 1,200 kW today, with machines up to 5,000 kW under development. Additional technological innovations, from lighter and more flexible blades to sophisticated computer controls, variable speed operation, and direct-drive generators, have driven operating costs down to the point where wind farms on good sites can generate electricity at an economically effective rate of 3–5 cents per kilowatt hour. Commercial wind power technological advances coupled with increases in natural gas prices have made wind power the least expensive source of new electricity in many regions. Meanwhile, the global wind power market is advancing rapidly. Installations increased from 1,290 MW in 1995 to 11,770 MW in 2005. (Energy Future Coalition 2008)

Private sector R&D dwarfs government investment in renewable wind energy, and the wind power industry is anxious to drive costs down even further. Global wind turbine manufacturing is dominated by companies based in the largest markets: Germany, Spain, and Denmark. The world's largest power-generation company, General Electric, entered the wind business and has become one of the world's top turbine producers. On the project development side, the U.S. market is dominated by a large, diversified power company, Florida Power and Light, which develops and owns wind farms.

Historically, the United States led the world in wind energy capacity, but abrupt changes in federal and state policies led to a collapse in the market. However, now a new federal tax credit, combined with an increasing number of supportive state policies, has led to a growing but sporadic wind energy market. Short-term extensions of the federal tax credit, often after long delays, have caused wild swings in new installations—from about 400 MW in 2002 and 2004, to approximately 1,700 MW of new capacity in 2001 and 2003—which have discouraged the industry from making long-term investments. Extension of the credit through 2007 helped drive another upswing in 2005: the United States installed a record 2,431 MW, adding more wind power capacity than any other country for the first time in over a decade. The empirical evidence points to the need for future tax credits to promote renewable wind power. (Interstate Renewable Energy Council 2008)

Wind farms were the country's second largest source of new generating capacity built in 2005. By the end of 2005, the U.S. Energy market nation had enough cumulative wind capacity to meet the needs of 2.3 million U.S. households. Looking at the European power markets gives another example of the success of wind power. In Denmark and some areas of Germany and Spain, wind meets more than 20 percent of electricity needs. The key to success in these countries is laws that provide renewable power producers with long-term power purchase agreements at prices sufficient to cover costs. This allows potential investors to enter into commercial projects with a minimized risk level. Instituting a consistent set of policies and by gradually lowering the purchase price as technology improves; European countries have nurtured a wind power industry that is already cost-competitive with new gas-fired power plants in most countries. (EREC2008)

Sun Power

Solar cells that convert sunlight directly into electricity are one of the most revolutionary new energy technologies to be commercialized in recent decades. Solar electric cells are adaptable to a remarkable range of uses, from handheld electronic devices to mountaintop weather stations, large desert power plants, and solar installations on rooftops. Commercial solar technology cells can produce electricity almost anywhere—the solar resource in Maine, for example, is about 75 percent of that in Los Angeles. (Solar Energy Industries Association 2008)

Solar cells were originally developed for use in orbiting satellites and were thought far too expensive for most earthbound energy applications. Improved manufacturing, efficiency gains, and economies of scale in production and installation of renewable solar technology have steadily lowered costs. Since 1976, prices have dropped by about 5 percent annually, and they continue to fall. New technologies under development, such as plastic solar cells, nanomaterials, and dye-sensitized solar cells, could enable the industry to leapfrog far beyond current technologies, further reducing costs while improving performance. As the technology becomes more affordable the attractiveness of commercialization increases. (EPA 2008)

Solar power is already the most economical way of providing electricity in many circumstances, particularly for small-scale devices like roadside call-boxes and off-grid telecommunications installations. Such uses are important but represent relatively small niche markets. Major market opportunities exist, however, for customers who value the security, power quality, and reliability that solar systems can provide—for emergency preparedness and security uses. Thousands of solar-powered homes have already been built in the United States—many of them in suburban neighborhoods, where excess power is fed into the electric grid, which later provides electricity for the home when the sun isn't shining. In southern California, builders and developers have begun promoting solar power as an inviting new feature. And elsewhere around the country, solar installations are appearing on high-rise apartment buildings, atop urban metro stations, and on the rooftops of rural businesses. (Florida Solar Energy Research Center 2008)

In some locations, rooftop solar power is now competitive with peak electricity prices. Commercial solar technological building materials can be cheaper than other façade materials, such as granite or marble, with the added benefit of producing power. A major drawback is that solar PV manufacturing requires hazardous materials, including many of the chemicals and heavy metals used in the semiconductor industry. However, there are techniques and equipment to reduce the environmental and safety risks and the industry is moving toward recycling of old solar cells.

The market potential for commercial solar power is enormous. Solargenix is constructing a 64 MW trough plant in Nevada that should be operational in early 2009. Stirling Energy Systems has signed power purchase agreements with two California utilities totaling 1,750 MW and plans to begin constructing a 1 MW pilot plant in California. Utilities in states with large solar resources (Arizona, California, Nevada, and New Mexico) are considering installation of solar dish systems as well. No commercial central receiver or tower plants have been built to date, but an 11 MW generator is under construction in Spain. According to the Western Governors' Association Solar Task Force report, within the next decade, 4,000 MW of central solar plants could be installed in the United States, generating thousands of new jobs. Alliance to Save Energy 2008)

The International Energy Agency estimates that total global installations of solar heating panels for all uses amount to about 196 million square yards, enough to cover the equivalent of more than 30,000 football fields A Department of Energy study projects that half of residential space heating and 65–75 percent of water heating needs could be met with solar. But stronger government support at the federal, state, and local levels will be needed if the United States is to keep up with the solar heating boom in other countries. (DOE 2008)

Hydropower

Hydropower uses the natural energy of falling and flowing water to produce electricity or mechanical energy. Historically, water wheels were widely used to grind grain and later to run America's factories until grid-connected electricity freed industrial processes to locate away from falling water. Currently, hydropower provides about one-fifth of the world's electricity and nearly 7 percent of U.S. power—the largest share of any renewable resource. In 2004, hydropower generated 270 billion kWh of electricity in the United States, a figure that has remained roughly constant for three decades. (National Renewable Energy Lab 2008)

Hydropower plants cost relatively little to run and can be operated and maintained by trained local staff. Hydropower installations generally have a long project life; equipment such as turbines can last 20–30 years, while concrete civil works can last a century or more. Unlike most fossil fuel fired power plants, the amount of electricity generated at hydro dams can be quickly increased or decreased, giving regions that have a large portion of hydro generation added flexibility in how they operate their power systems. Hydropower can help maintain grid stability and can be called up when other power sources fail. Flexibility allows for a sizable share of intermittent renewable capacity from solar or wind energy—which can be easily backed up with hydropower. (National Hydropower Association 2008)

In principle, U.S. hydropower generation could be increased significantly. The Department of Energy (DOE) reports that hydropower could double its current contribution of more than 78,000 MW. According to DOE, 21,000 MW of capacity could be added simply by improving existing projects and installing generators at dams that do not have them. Of the

80,000 dams in the United States, only 3 percent are used to generate electricity. Despite this potential, the industry has experienced sluggish growth over the past decade. As with other renewables, upfront capital costs are high. The licensing process can be time consuming and costly and the lack of tax incentives for hydropower has served as a disincentive to growth. Hydropower can be made a more attractive option by providing financial incentives and increasing energy efficiency. (National Hydropower Association 2008)

The vast majority of the nation's hydropower comes from large-scale facilities, but a significant share of U.S. hydro plants today are micro-scale (up to 100 kW) or small-scale systems (100 kW to 30 MW). Rather than using a large dam and storage reservoir, micro- and small-scale projects generally use "run-of-river" designs that produce electricity by diverting only part of a stream. Most consist of small turbines that rely on water pressure or velocity to generate power. Small hydro facilities often have difficulty gaining affordable grid connections, and power purchase agreements with utilities are generally required for independent power producers to operate such systems. (Kunstler 2005)

Critics argue that habitat alteration, disruption of fish migrations, trapping of sediment, displacement of communities, and greenhouse gas emissions from rotting organic material are among the possibly irreversible impacts of hydropower. The industry is pursuing a variety of measures to reduce such impacts. And even small hydro is hindered by the perception that it can adversely affect fishing. But environmental impacts can be curtailed through good system design and appropriate construction and operating practices. Small-scale hydro systems cause little change in stream channel and flow, and thus have minimal impact on water quality, fish migration, and surrounding habitat. (National Hydropower Association 2008)

Micro Power: A Paradigm Shift

Although most of today's electricity comes from large, central-station power plants, new technologies offer a range of options for generating electricity where it is needed, saving on the cost of transmitting and distributing power and improving the overall efficiency and reliability of the system. These new options include renewable energy technologies such as rooftop solar cells and fuel cells that may run on energy sources derived from sources other than fossil fuels. Micro power is in effect a return to the vision of Thomas Edison, who designed small, city-based power plants, the first of which was built near Wall Street in 1882. Economies of scale quickly rendered this approach obsolete, but new technologies that can be mass-produced at low cost are bringing us back. Locally based generators that connect to local distribution lines generally have generating capacities of 5 MW or less, and are sited in or adjacent to residential, commercial, or public buildings. (Brown 2001)

Micro power plants provide additional value to the electricity system because they do not require extra investment in transmission or distribution, and they reduce or eliminate transmission line loss. The popularity of micro power plants is also fueled by the need for reliable power supplies for the electronic equipment that is so central to today's economy. Since most power outages are caused by weather-related damage to power lines, locally based generators can dramatically improve reliability. Japanese companies have demonstrated that the development of simple, integrated technology packages can quickly and significantly reduce the cost of home-sized solar generators. Recently, U.S. companies have introduced so-called "plug-and-play" solar systems that are modular and elegant—easily integrated into a new or existing building without the need for custom design work. Solar experts believe that as these systems become more standardized, commercial and residential consumers will see the units proliferating in their neighborhoods over the next few years. (Union Of concerned Scientists 2008)

An empirical example of micro power distribution can be found in California. One business that has taken advantage of small-scale solar power is the FedEx Corporation. In 2005, FedEx completed a solar electric system atop its hub at Oakland International Airport. The 81,000-square-foot system generates enough electricity to power 900 homes, and provides 80 percent of the facility's peak load while protecting the roof from UV rays and reducing heating and cooling needs. (Center for Resource Solutions 2008)

The fact that micro generators are not widely used today reflects in part the fact that everything from electricity laws to environmental and tax regulations are often structured in ways that disadvantage these technologies. Despite such impediments, businesses and consumers increasingly demand the ability to generate their own power and to sell electricity to other consumers at a fair price. Under "net-metering" laws that have been enacted in several states, it is now possible for consumers to sell some of their extra power back to the grid at the same price the consumer pays for it. These laws have helped spur the growing popularity of rooftop solar power systems, particularly in California. (Center for Resource Solution 2008)

Sustainable Land Use

Renewable energy is commonly viewed as too land-intensive to be practical. Yet harnessing renewable energy requires less land and water than does our current fossil fuel based energy system. Disputes over the location of renewable energy projects—particularly wind farms, such as the Cape Wind project off the Massachusetts coast—are not uncommon. Solid regulatory procedures and strong public participation can ensure that a balance is struck between energy production and environmental and aesthetic considerations.

Studies show that wind resources in three states—Kansas, North Dakota and Texas—could in principle meet all current U.S. electricity needs. Although wind farms appear to occupy as much as 60 acres per megawatt, depending on the terrain, the turbines and access roads actually cover under three acres per megawatt. By conservative estimates, this means that fewer than 1,400 acres are needed to produce one billion kilowatt-hours (kWh) of electricity each year. Farming and grazing can continue beneath the wind turbines, enabling farmers and ranchers to supplement their incomes with payments for green power production. (Utility Wind Integration Group 2008)

Moreover, the Great Plains, where most of the best wind resource is located, is one of the least densely populated parts of the country. Geothermal electricity is estimated to need just 74 acres of land to generate one billion kWh of electricity annually, enough to power nearly 94,000 American homes. By contrast, coal-fired power requires 900 acres per billion kWh generated annually—most of it for mining and waste disposal. The geothermal plant can go on producing electricity on the same land for a century or more, as can wind farms, while a coal plant depends on mining hundreds of additional acres each year. (Utility Wind Integration Group 2008)

Solar power plants that concentrate sunlight in desert areas require 2,540 acres per billion kWh. On a lifecycle basis, this is less land than a comparable coal or hydropower plant requires, and because most deserts are sparsely populated, there is plenty of room for solar power plants. A little over 4,000 square miles—equivalent to 3.4 percent of the land in New Mexico—would be sufficient to produce 30 percent of the country's electricity. Additionally, sunlight can be used to produce power without using any land at all, simply by installing solar cells on the available roofs and walls of U.S. buildings. It is estimated that the nation has 6,270 square miles of roof area and 2,350 square miles of façades that are suitable for harnessing solar power. Mounting solar panels on just half of this area could supply nearly 30% of U.S. electricity. (Solar Energy Industries Association 2008)

Solar and wind power require virtually no water to operate as opposed to large fossil and nuclear plants, which, need enormous quantities of water for cooling and ongoing maintenance. According to the Union of Concerned Scientists, a typical 500-MW coal plant takes in 2.2 billion gallons of water—enough for a city of 250,000 people—each year simply to produce steam to drive its turbines. Crops grown for biofuels are the most land- and water-intensive of the renewable energy sources. In 2005, about 12 percent of the nation's corn crop (covering 11 million acres of farmland) was used to produce four billion gallons of ethanol—which equates to about 2 percent of annual U.S. gasoline consumption. For bioenergy to make a much larger contribution to the energy economy, the industry will have to accelerate the development of new feedstocks, agricultural practices, and technologies that are more land and water efficient. Already, the efficiency of biofuels production has increased significantly. (American Coalition on Ethanol 2008)

Bio Power and Bio Mass Fuels

The same homegrown resources that can fuel America's vehicles can heat and power our industries, businesses, and homes. Biopower is the process of using organic matter from America's fields, forests, and landfills to generate electricity. It is the nation's largest non-hydropower source of renewable electricity. Biopower currently provides only about 2 percent of U.S. electricity, but it has the potential to meet a much larger share of power demand while reducing pollution and revitalizing rural communities. America's biomass resources range from agricultural and forestry residues, to animal waste, to fast-growing plants grown solely for energy production. Landfills can also be tapped, by capturing methane from biodegrading organic wastes before it escapes to the atmosphere. (Biomass Council 2008)

Biomass is extremely versatile, it can be burned directly to produce steam, which turns a turbine to generate power; it can be co-fired with fossil fuels; and it can be gasified to produce steam and electricity, or for use in micro turbines or fuel cells. Most biopower is used by the forest products industries, which produce steam and power with process residues. More than 100 U.S. coal-fired power plants are now burning biomass together with coal. Experience has shown that biomass can be substituted for up to 2–5 percent of coal at very low incremental cost; higher rates—up to 15 percent biomass—are possible with moderate plant upgrades. According to the Washington Department of Ecology, the state produces enough biomass to generate over 15.5 billion kWh of electricity, or almost half of Washington's residential power consumption. (Biomass Council 2008)

Growing energy crops for biopower poses the same environmental concerns associated with biofuels. Burning biomass in power plants releases particles that can affect human health, as fossil fuel burning does, but pollution control technologies can remove these particles from the smokestack. When burned with coal, biomass can significantly reduce emissions of sulfur dioxide, carbon dioxide (CO_2), and other greenhouse gases. Burning biomass destined for landfills also reduces the amount of organic waste that would ultimately decompose and release methane, a green house gas that is 21 times more potent than CO_2. . (Biomass Research and Development Initiative 2008)

Capturing methane from the decomposition of organic matter found in landfills, sewage treatment plants, and livestock facilities provides premium fuel while reducing the amount of waste that must be disposed of. Using anaerobic digesters at all U.S. farms where they would be economical could avoid emission of an estimated 426,000 metric tons of methane annually. This practice is starting to catch hold in large hog, poultry, and cattle operations, driven by the need to control emissions and by the lure of selling lucrative energy. Central Vermont Public Service sells electricity produced from farm waste directly to consumers, and will soon generate enough power for 1,400 Vermont homes. (Biomass Research and Development Initiative 2008)

Biopower can provide baseload electricity, and plants can be located close to the point of demand, reducing the need for expensive upgrades to the power grid and minimizing transmission losses. Additionally, biopower can generate up to 20 times more local jobs than natural gas-fired power plants do. Facilities can range in size from small farm-based operations to much larger plants. As with other renewable technologies, inconsistent availability of subsidies has hampered industry development. In addition, the permitting process is often time-consuming and expensive, and a lack of national grid connection standards often complicates development. These policies must be reformed if biopower is to fulfill its promise. The government must provide financial incentives and subsidies in order to further the growth of the biomass energy industry.

Energy Efficiency the Cornerstone of Green Economics

One of the main problems associated with fossil fuels is the economic impact of this type of energy. The destructive impact of fossil fuels can be witnessed in a broad range of economic sectors. Improving energy efficiency represents the most immediate and often the most cost effective way to reduce oil dependence, improve energy security, and reduce the health and environmental impact of our energy system. By reducing the total energy requirements of the U.S. economy, improved energy efficiency will make increased reliance on renewable energy sources more practical and affordable. Historically, energy efficiency has played a critical role in the U.S. energy supply, reducing total energy use per dollar of gross national product (GNP) by 49 percent since the 1970s. Compared to a 1973 baseline, America now saves more energy than it produces from any single source, including oil. Efficiency improvements stabilize energy prices by reducing demand, while also delivering the same services we value—whether hot showers or cold drinks—at lower cost. Energy efficiency is effective at promoting energy independence and cost effective supplies of reliable power. (Pew Center for Climate Change 2008)

The potential for additional energy savings is vast: U.S. energy use per dollar of GNP is nearly double that of other industrial countries. More than two-thirds of the fossil fuels consumed are lost as waste heat—in power plants and motor vehicles. The fuel economy of new U.S. motor vehicles advanced rapidly, from 14 miles per gallon in the mid-1970s to 21 miles per gallon in 1982, driven by rising fuel prices and government- mandated fuel economy standards. But in 2006, new U.S. vehicles still averaged just 21 miles per gallon; for over two decades, automakers have put most of their engineering efforts into building larger vehicles with more powerful engines, offsetting the potential fuel economy gains from new technologies. (Energy Efficiency and Renewable Energy, DOE 2008)

The time is ripe for another great leap in vehicle efficiency. New technologies such as hybrid drive trains, clean-burning diesel engines, continuously variable transmissions, and lightweight materials could allow vehicle fuel economy to double over the next two decades. Significant efficiency gains are also possible in the electricity sector. Americans spend $200 billion annually on electricity, but current demand could be halved with cost-effective technologies already available on the market. Furthermore, decreasing electricity demand reduces the need for new, large power plants, allowing smaller, distributed, renewable generation to play a greater role in meeting our energy needs. Past experience demonstrates that strong government policies for advocacy and promotion can spur the private sector to invest in efficiency improvements.

Since national home appliance efficiency standards were enacted in 1987, manufacturers have achieved major savings in appliance energy use. Refrigerator efficiency nearly tripled between 1972 and 1999, and dishwasher efficiency has more than doubled in the last eight years. California's *Flex Your Power* campaign, enacted in response to the state's 2001 energy crisis, immediately reduced power demand by 5,000 megawatts by replacing millions of standard light bulbs with compact fluorescent lights (CFLs), installing light-emitting diode (LED) traffic lights, and replacing inefficient appliances. With the implementation of robust efficiency policies, California has the lowest per capita energy consumption in the nation, without sacrificing comfort or valued services. (Data Energy Efficiency Council 2008)

Technologies available today could increase appliance efficiency by at least an additional 33 percent over the next decade, and further improvements in dryers, televisions, lighting, and standby power consumption could avoid more than half of the projected growth in demand in the industrial world by 2030. The integration of efficiency with renewable energy maximizes the benefits of both. For example, the correct building orientation can save up to 20 percent of heating costs; those savings can jump to 75 percent when renewable energy and appropriate insulation are integrated into the building. A national commitment to improved efficiency can transition the U.S. energy economy in ways that will yield dividends for all Americans.

Dependence on Foreign Oil

Present energy policies jeopardize economic health and well being by increasing United States dependence on foreign sources of oil. Relying on fossil fuels for power for the US economy puts our country in a precarious position by focusing reliance for imported oil in the hands of a few Middle Eastern countries. Energy security and the long term viability of energy policy are put in jeopardy by the country's continuing reliance on oil imports. (Worldwatch Institute 2008)

America's dependence on imported oil is undermining the country's national security by tying the U.S. economy to unstable and undemocratic nations, thus increasing the risk of military conflict in political hotspots around the globe. Renewable energy can reduce oil dependence and improve the country's security in several key ways. The United States currently imports some 13 million barrels of oil each day—over 60 percent of its total daily consumption—at an annual cost of $300 billion. If current trends continue, America will depend on imports for 70 percent of its oil by 2025.

As President Bush said in his 2006 State of the Union address, America is "addicted to oil." This addiction requires billions of dollars in military expenditures to secure the country's energy supply lines. As accessible reserves in the world's stable regions have been depleted, oil extraction has gradually shifted to more dangerous corners of the globe. Today, the world's oil frontier includes a list of countries that mirrors a catalog of global trouble spots, including Angola, Azerbaijan, Chad, Nigeria, Sudan, and Venezuela. Most of these oil exporting countries rank disturbingly low in many measures of political liberty, human rights, and corruption. An estimated 85 percent of the world's oil reserves are now either owned or controlled by national petroleum companies, which greatly limits private investment in exploration and infrastructure development. (Energy Future Coalition 2008)

The Middle East contains a remarkable 60 percent of the world's remaining proven oil reserves, and each day, nearly half the world's oil exports travel through the Straits of Hormuz at the mouth of the Persian Gulf. Because of their geographical proximity, Europe and Asia import a larger share of their oil from the Middle East than the United States does. But this does not lessen the U.S. exposure to imported oil. For three decades, the Middle East has been the world's marginal oil supplier, and disruptions in the flow of oil are reflected in the world price of energy and the balance of global economic power. In recent years, however, even the large oil reserves in the Persian Gulf have been insufficient to keep up with rising global demand, most of it coming from the United States, the Middle East, China, and other Asian countries. If supply fails to keep up with rising demand, oil prices could rise far above their recent record highs. Every oil price spike over the past 30 years has led to an economic recession in the United States; such price spikes will become more frequent as global competition for remaining oil supplies intensifies. (IEA 2008)

Full U.S. energy independence will take decades to achieve; until then, national security could be greatly improved if the economy moved from its current path of rising oil imports to reducing national reliance on oil. Reducing demand for foreign oil is an eminently achievable goal— through both transportation efficiency improvements and increased reliance on biofuels and other renewable resources. Improving efficiency and diversifying fuel choices will take the pressure off energy prices, while enabling the country to make diplomatic and security decisions based on American interests and values rather than the relentless need to protect access to oil. In many areas of the world, the U.S. diplomatic hand would be greatly strengthened if energy imports were going down rather than up.

The current energy system undermines national security in other ways as well. The centralized and geographically concentrated nature of the country's power plants, refineries, pipelines, and other infrastructure leaves it vulnerable to everything from natural disasters to terrorist attacks. One year after Hurricane Katrina crippled approximately 10 percent of the nation's oil refining capacity, oil and gas production and transportation in the Gulf of Mexico still have not been fully restored. Security experts believe that a well-orchestrated physical or electronic attack on the U.S. electricity grid could cripple the economy for an extended period. It is estimated that the 2003 Northeast blackout cost between $4 billion and $10 billion over the course of just a few days. (American Council for Energy Efficient Economy 2008)

The country's 104 nuclear power plants and their associated pools of high-level radioactive waste present another U.S. security threat. If one of the planes that struck the World Trade Center on September 11, 2001, had instead hit the Indian Point nuclear plant just north of New York City, the human and economic toll of that fateful day could have been vastly greater.

The distributed nature of many renewable energy technologies helps reduce the risk of accidental or premeditated grid failures cascading out of control. An analysis of the 2003 Northeast blackout suggests that solar power generation representing just a small percentage of peak load and located at key spots in the region would have significantly reduced the extent of the power outages. (Union of Concerned Scientists 2008)

A 2005 study by the U.S. Department of Defense found that renewable energy can enhance the military's mission, providing flexible, reliable, and secure electricity supplies for many installations and generating power for perimeter security devices at remote installations. Renewable energy provided more than 8 percent of all electricity for U.S. military installations by the end of 2005. Both the military and the Central Intelligence Agency are turning to new lightweight solar technologies to replace heavy batteries in the field and for use in intelligence applications. (Union of Concerned Scientists 2008)

Renewable energy can play an important role in providing power to critical infrastructure in the aftermath of catastrophes as well. For example, the Louisiana State Police used solar powered lighting in critical areas around New Orleans following Hurricane Katrina; elsewhere in Louisiana, the lack of power slowed the work of emergency and recovery workers. Officials at New Jersey's Atlantic County Utilities Authority plan to install solar and wind power at a waste-water facility to keep the plant operating during blackouts. (Interstate Renewable Energy Council 2008)

Renewable technologies can be coupled with traditional backup diesel generators to extend the fuel supply and increase the total power available. Renewable power can also come back on line much more quickly than coal or nuclear power plants can, which helps to reduce economic losses associated with power failures and minimize the time that critical facilities such as hospitals and emergency communication centers must go without power. As with oil dependence, the broader energy security threats cannot be eliminated overnight. But immediate steps to invest in a diverse, decentralized energy system that relies more heavily on domestic renewable resources will allow the United States to steadily enhance its security.

Chapter Three

U.S. Energy Policy and Statistical Methodologies

This chapter will address in five separate areas the research necessary in order to make a reasoned judgment concerning the future direction of US energy policy. The chapter will specifically:

1. Outline the collection of data

2. Specify research parameters

3. State data collection steps and procedures

4. Summarize the analytical findings

5. Provide validity with regards to the measurement tools.

Overview

There will be three key variables or rather benchmarks (cost, technological feasibility and the ability to provide power) which will be used to measure the success or credibility of alternate energy sources. The alternatives (solar wind and tidal power along with energy efficiency will be judged against three key criteria in order to determine their effectiveness when examined vis-a-vis the other traditional fossil fuel energy options. The analysis will examine **energy provision,** that is the ability of the alternative in question to adequately provide power per kilowatt hour. These criteria will be examined using cost benefit analysis to determine the most effective providers of energy. Second, the research will examine the **financial cost** of the proposed energy alternative. The financial impact of energy sources will be used to evaluate and determine their overall effectiveness. Last **technological feasibility** will be considered as an option to determine the feasibility of the proposed alternative. These three variables will drive the focus of the research and will determine the boundaries for the research. The variables will allow professionals to determine which energy choice is the most germane given their surrounding circumstances and needs. Specifically, the placement of these energy sources in the construction, agriculture and transportation sectors will be evaluated.

The ultimate goal of the research effort is to establish which sources of energy are the most reliable and dependable when compared side by side. The research will seek to explore patterns or themes in the data concerning the operation of energy technologies. The ultimate goal is to provide insight into existing findings while paving the way for future research efforts. The paper is a snapshot and will provide valuable insight into the present body of literature pertaining to alternative energy sources. The research will ultimately answer the why, the how, and what of energy policy. The research sets out to study more in depth the social and cultural impacts associated with energy policy. A thorough analysis will be undertaken to provide a qualitative assessment of the present literature as it concerns energy efficiency and renewable energy technology.

A number of sources will be investigated in order to establish reliable data validity and to provide insight into free market economics and U.S. energy policy. Specifically, the research will provide an assessment of energy efficiency and renewable energy technology as they relate to economics, the environment and technological commercialization efforts. The research will be analyzed and conclusions will be formed based on triangulation of data from all relevant sources. The methodology is thorough in is assessment of relevant literature and will form summary judgments based on information which is gathered from a variety of sources. The study and research will try to present a preponderance of evidence to demonstrate that each of the conclusions are the most reasoned judgment to be conferred given the examination of the pertinent data. Specific assessments will be made for solar, tidal, wind and energy efficiency.

The effort of this research is to answer research questions and fill in gaps in existing bodies of knowledge as they relate to energy distribution and potential energy savings. This effort will objectively evaluate non fossil fuel sources of energy and determine whether or not these sources are capable of providing enough energy to consider them effective alternatives to existing sources of energy. This guided and deliberate research effort will also provide additional insight into the economics of alternate sources of energy. Data (theoretical and empirical) will be analyzed to determine which non fossil fuel can provide cost efficient energy delivery in today's completive energy marketplace. The research will look at the technological feasibility of transitioning away from fossil fuels. Research will demonstrate the practicality and financial attractiveness of alternate energy sources and energy conservation.

The research is aimed at a through and exhaustive analysis to determine the existence of any patterns or themes in the existing literature surrounding alternative fuels. Specifically, the focus will be on conservation and energy efficiency technologies and alternative sources of power --wind solar and tidal. The analysis seeks to plug holes in the existing body of knowledge underlying energy efficiency and alternate fuel technology.

The research addresses the primary question of why the U.S. government has yet to fully embrace energy efficiency technology and promote alternate sources of energy. How the government and the private sector can react to and benefit from the energy marketplace will be explored in detail. What changes the government needs to make in policy and what efforts are needed from the private sector will be evaluated. The steps required to institute the most cost efficient, technologically feasible and effective energy production technologies, will be investigated and recommended to become a part of U.S. energy policy. The study is of vital importance given its timing and context--the present state of energy policy remains in flux and global demand for energy is at an all time high. Despite proclamations to the contrary, the U.S. government does not effectively promote conservation technologies and continues to chase oil as a means of global energy policy.

Benefits of Methodology

The benefits of the qualitive approach are numerous and this methodology is clearly the most appropriate when assessing energy policy. In much of the literature, there are no clearly identifiable benchmarks to evaluate energy policy. This analysis identifies three easily recognizable variables (cost feasibility and production) against which the three forms of energy (solar wind and tidal power) can clearly be assessed. This book represents a comprehensive focused effort which will effectively evaluate and address the shortcomings in the present state of research.

This analysis and summary will allow policymakers to make informed decisions regarding which energy sources are the most beneficial technologically, economically, and production wise-as alternate sources of energy production. The analysis will provide a solid foundation of knowledge which can be used to refine existing policy to enhance the effectiveness of existing energy policy, and pave the way for a bold new energy policy which does not rely on fossil fuels as its cornerstone. The knowledge produced can also be infused into the formulation of new, future energy policy. The central collection and condensation of the body of information pertaining to energy efficiency conservation and alternative energy technologies can be a valuable tool in shaping the future direction of U.S. and global energy markets.

Specifically, the use of qualitive methods will allow new paradigms to emerge based on yet undiscovered data and trends. The collection and assimilation of data will in most cases lead to the development of new paradigms and modes of thinking. This methodology will allow a more sustainable energy paradigm to emerge with conservation and alternate energy at the forefront. This is important for a paradigm shift will be required to transition away from a fossil fuel based economy.

This methodology will also uncover any hidden variables which may be influencing the data or the trend analysis. The consolidation, triangulation, and evaluation of data will allow researchers a bird's eye view of the present state of energy policy. This effort will cause extraneous variables to be eliminated and will hone in and focus on those variables which are critical to making and effective determination for the future direction of energy policy. By focusing on and poring over those variables which are most important (cost, technology and production) researchers will be able to make more reasoned and educated policy decisions regarding the inclusion/exclusion of these variables.

Additionally this effort will allow the formulation of theories where currently none exist. A new way of thinking and new theories concerning energy policy are exactly what is necessary if the government is to pursue an energy policy which is not married to fossil fuel technology. It is this type of information generation that is perhaps the most appealing benefit of this research. This body of knowledge can be used to encourage bold new thinking and the institution of measures which have heretofore been avoided as a part of energy policy. The provision of knowledge on solar, tidal, and wind power may be the necessary spark to initiate a change in the direction of energy policy.

Lastly, the analysis of quality of data will allow the removal of researcher bias. In this case it of the utmost importance that researcher bias is drained out of the analysis. In many instances data is provided by companies who have a vested interest in the success or failure or the technology in question. A third party is able to neutrally observe and evaluate the data which gives one a clear glimpse in to the true attractiveness of the energy technology being evaluated. The slate is wiped clean and a thorough unbiased analysis is the result. The technologies will be assessed across three markets-agriculture transportation and construction. The variables will be overlaid in a matrix allowing the examination of each given technology in each of these three markets. Researchers will be able to evaluate wind power, solar energy, tidal energy and energy efficiency measures across all three market segments.

The methodologies employed are considered best practices for energy policy evaluation. The approach provides a real world examination with practical implications which can be immediately translated into a successful energy policy. There are some key differences which segregate this approach from those employed in traditional quantities analysis. In this instance there is not a specific hypothesis to be tested. There is not a pure scientific laboratory approach which will either accept or reject the hypothesis; rather there is an intense research effort which will produce tangible results which can be immediately translated into a more effective energy policy.

Qualitative Analysis

The analysis is in contrast to numerical quantity based investigations; there is no effort to emulate the findings of previous researchers and add to their validity, rather there is an effort to learn from the existing knowledge and add to that data in order to further its effectiveness and validity. While the statistical analysis is not the main thrust of this research it still remains a central component for it is nigh impossible to evaluate energy production efficiency and cost schedules without mathematical formulas. This effort seeks to explore patterns which are based primarily on units of measurement that are terms not numbers. At the end of the research any interested observer will be able to determine the most effective energy technology to be employed in any of the sectors (construction, agriculture, and transportation) targeted by the research.

In many instances the discovery or endorsement of given energy technologies can have a very immediate and pronounced effect on energy markets. The employment of the technology by the federal government and the existence of subsidy programs can be the difference between a technology that flourishes and one which falters. As new energy choices are presented a mindset or paradigm will need to be embraced which allows for the introduction of alternate fuel sources. New paradigms based on the trends and findings can be discovered and/or implemented. A new way of thinking will need to be instilled which automatically gravitates the choice for energy towards those sources which are the most cost effective, technologically feasible and capable of providing reliable sustainable sources of energy.

Qualitative designs such as those utilized in this assessment help researchers explore patterns or themes seeking to establish a conclusion. The research conducted herein is offered as knowledge which can provide valuable information to fill in gaps existing in present theoretical research. The research assessment and the literature review are set forth in order to discover the why and how of energy policy. The assessment will attempt to identify economic drivers of energy efficiency and renewable energy technology in order to maximize investment on potential return. The methodology is one of research comparison and qualitive analysis to determine and evaluate the effectiveness of germane energy technologies.

The purpose of this study is to shed light on existing theories and schools of thought. This research will serve as a roadmap for future research and scholarly discussion. The paper relies on the popular constitutional notion of the marketplace of ideas. This theory is often used when discussing first amendment freedoms to show how a healthy discussion is most likely generated by allowing all voices to be heard and forcing all theories to be held accountable. A vigorous healthy discussion will in almost all instances universally yield a reasoned judgment. A healthy thorough literature review will allow great insight into the realm of energy policy.

Methodical research is used to illuminate popular schools of thought in the field of energy policy. The debate will be explained in terms which will allow a greater understanding of energy efficiency technology and renewable energy technology. By looking at present alternate fuel installations or renewable and energy efficiency technologies and utilizing expert testimony this report will uncover some of the nuances surrounding the technologies. Research will allow us to extrapolate present results onto a larger scale. The knowledge gained through the literature review will give policymakers required information to promote technologies which will be the most environmentally responsible and economically effective.

By using expert testimony and computer modeling we will also be able to address some of the short and long term effects of renewable and efficiency technologies. Examing the empirical success of these installations will allow a more successful energy policy which is less environmentally damaging and more sustainable over the long term. . The knowledge base will allow the most efficient technologies to be discovered and promoted in order to fully realize their benefits. The methodology will allow researchers to understand the impact these technologies may play in pollution abetment and as a tool to combat global climate change. Statistical review will be sufficient but not necessary to address many of the questions presented here. What is needed is a fundamental assessment and rethinking concerning US energy policy.

Although a widespread review of energy policy is beyond the scope of this research, the review of data will answer the question of whether or not the US economy can be powered with alternate non fossil fuel based economies. The research will also allow us to glimpse into an economy which is not run by fossil fuel technology but instead relies on the more efficient use of existing technology and renewable energy sources. The end result will ultimately lie in the hands of decision makers and policymakers who will need to champion the technologies are the most desirable and historically most effective. This investigation will allow us to fill in presently existing gaps in knowledge so that the best energy efficiency and renewable technologies are adopted. The research is aimed at discovering the why, how, and what of the things that are occurring in the area of energy policy regarding fossil fuels and energy generation.

The research will look for new opportunities and innovative possibilities based on the research that is uncovered. New paradigms and ways of thinking will be analyzed in order to make the best decisions regarding future research and commercial applications of this technology. The least costly and most effective technologies will be reported in order to make energy policy economically cost effective and environmentally desirable. The paper will examine the research problems in new ways, paying particular attention to the importance of the economic effects of the alternate energy technologies. The research will reveal the most effective reliable ways of providing power in a sustainable manner. Ultimately the knowledge gained will allow the promotion of new theories which guide future efforts at energy efficiency and commercialization of renewable technologies. This paper attempts to describe and interpret data with the goal of detailed and well-rounded results the conclusion of these findings should serve as a springboard for future efforts at energy sustainability.

There are a number of benefits to using the approach favored by this research effort to evaluate the reliability and applicability of alternate energy sources. First, new paradigms may be established as a result of the research findings. The new paradigms may be fueled or supported by the emergence of trends in the data. This seems particularly true in this body of work. For nowhere else is the emergence new paradigm more likely than the energy sector. The possibility exists for an energy future much different from the present reliance on fossil fuels.

Triangulation

One of the key methods involved in this research is triangulation. This analysis looks at evidence from multiple sources in order to greater enhance data reliability. This method will eliminate some problems which normally occur in research such as identifying researcher biases and assumptions. It is very important that data is triangulated and understood from a variety of perspectives; particularly when looking at commercial applications it is critical to learn from the historical experience these installations provide. Triangulation in this case means combining results from different sources to increase confidence in its reliability and accuracy of results. Triangulation in qualitative designs means that the data originates from many sources such as archival files, interviews, articles, observations; replicating patterns in variety of sources increases the reliability of qualitative studies. This research will first focus on the renewable energy options under consideration- wind, solar and tidal power; the discussion will then turn to a look at energy efficiency measures and the impact these moves may have on the economy and the environment.

Variables

The key distinctions which will be examined are the ability of non fossil fuel sources to compete with traditional fossil fuels sources of energy. Specifically, these sources will be compared and judged versus one another based on three key variables across three major market segments. The research will focus on economic cost of energy provision, the technological ability to consistently provide a reliable source of power, and lastly the output or amount of power generated by the sources under examination. These three criteria are the most salient and most critical factors when deciding on what type of power to use to provide power especially in distributed generation applications.

The three markets in question construction, transportation and agriculture are the biggest consumers of energy in the global marketplace and the most important markets for energy installations. When studying energy alternative choices it is imperative that these major market segments are given consideration and research addresses which types of power should be employed in these vital segments. During the examination of which powers sources are most logical for placement in these markets it is crucial that researchers and policymakers understand the cost, the technology, and the power production of the sources in question. The units of measurement will be scaled according to the particular power source or power segment under consideration. By honing the research in on these variables and market segments individuals will gain a much clearer insight into the applicability and relevance of these energy provision technologies.

Technology and production as the key variables when evaluating energy efficiency, solar, wind and tidal power sources. The research will look at power production across the three major market segments-agriculture, construction and transportation. The research will focus on finding common patterns and themes which are present in the literature and making these themes relevant to the study at hand. The research will examine the findings of past research and integrate these findings into the present research of this paper. The specific choices considering research parameters and guidelines for this research are set up with future research applications in mind. The format is designed to remain timely and helpful in guiding future research efforts and aiding policymakers as energy choices and policy decisions are forged in the future. The results of this research will provide an immediate roadmap to energy policy and decision makers so that the findings may be immediately incorporated into an effective energy policy at the federal and state government levels.

The following table is an example of the matrix type evaluation which will be used in order to facilitate and effective evaluation of the power sources using a table to provide the necessary information.

Solar Power -Construction Market

Cost	Technology	Power Provision
Cost per hour data	Applicability /installation	Watts generated

Tables will be generated for each type of power being considered and for each market segment under investigation. These tables will provide an easy succinct summary of the research findings that can easily be digested and used by policy and decision makers in energy markets around the world. Researchers will be able to easily determine which technologies are most appropriate for which markets.

The research for this paper comes from a variety of industry sources and expert testimony. The research also involves the evaluation and critique of hundreds of distributed generation sites which have been visited by the lead researcher. The variables were chosen based on knowledge of the industry and relevance to the study and long term energy policy.

The data collection comes from a number of sources in order to provide increased reliability to the data. The triangulation of the data is quite useful in improving the quality and accuracy of the data. The combination of sources, from the government, the private sector and from expert testimony has produced a very reliable insightful glimpse into the energy industry of the United States. The data originates from many sources such as interviews, on site visits, and literature reviews. The diversity of sources and the occurrence of patterns and themes over and over throughout the findings adds additional credibility and weight to the findings.

After the accumulation of this data the analysis will occur to produce a comprehensive in depth view of the American energy industry. Specific recommendations will be made concerning the further introduction of alternate power sources into the construction agricultural and transportation markets. The analysis will take place with energy sources highlighted and grouped by their units of measurement. The initial categories will be evaluated in order to determine the future applicability of these types of powers sources (solar, wind and tidal).

The results will be explained and grouped by variables (cost, feasibility and production) in Chapter Four in order to highlight the relevance of these findings. The results will be laid out in order to facilitate further research in to provide policymakers with real world data and numbers to facilitate the rapid introduction of alternate fuel technologies into the markets for which they are the most necessary and relevant. Chapter Five will summarize the findings in order to provide comprehensive insight into the energy markets. The "who cares? "questions will be addressed, as the data is translated in to usable formats to provide help to concerned policymakers. The policy implications will be addressed and additional outlier responses will be analyzed and incorporated into the summary.

Grounded Theory

The promotion of energy efficiency in the short term coupled with a long term transition towards alternate energy sources water wind and solar will effectively displace the need for fossil fuels in the free market. This is the underlying theory motivating this research. As a result of over twenty years experience in the industry and tireless research, the findings will result in one of the most comprehensive objective efforts into the commercial applications of alternate fuels. The initial parameters of this research were designed with the afore mentioned grounded theory guiding them. The idea was to qualitatively "test" the idea that alternate sources of power coupled with energy efficiency improvements could essentially displace the need for fossil fuels.

The theory is developed from the raw data surrounding alternate energy technologies. The initial pattern of this analysis guided by the formulation of a "grounded" theory. The initial parameters for this research are grounded in theory. This was done for two reasons. First, it allowed the study to have a very specific and relevant purpose. Second, it was conclusion based; that is the parameters were set up by definition to enhance the relevance and viability of the study. Grounded theory is particularly useful when a past perspective could generate new theories for the future framework for policymakers ad decision makers. Grounded theory is developed over time based on knowledge of the industry and an analysis of trends and technologies within the give field of study-such as the energy industry.

There are a number of qualified energy alternatives available which are technologically capable of serving the present power needs. As we look at the long term requirements of U.S. energy needs it is clear that we need to undertake a serious reassessment of what our priorities are and what the foundations for our policy should be as we move forward. Isolating key variables (cost feasibility and production) when researching specific alternate energy technologies (solar, tidal and wind) will provide information to guide energy policy and allow the implementation of alternate energy sources in the major markets of transportation, construction and agriculture.

Chapter Four Data Assessment

Alternatives to an Oil Based Economy

This chapter will provide outcomes for the research as it pertains to the relevance of renewable and alternate energy technologies. Specifically, the research examines each energy source (Solar, Wind, Water and Energy Conservation) as they relate to the markets (Construction, Agriculture and Transportation) being researched and evaluated.

Overview of Data

The Global Marketplace renewable energy is rapidly becoming big business around the world. Between the mid-1990s and 2005, annual global investments in "new" renewable energy technologies rose from $6.4 billion to $38 billion. It is estimated that investment in renewable energy technology could approach $70 billion by 2010. Wind and solar power are the world's fastest growing energy sources today, with capacity expanding at double-digit rates every year over the past decade. Other sources are growing rapidly as well, at rates far outpacing those for traditional energy sources. The global power industry is now adding more wind energy generating capacity to the world's grids each year than it is nuclear capacity. (European Union New and Renewable Energies 2008)

The effects of such rapid growth include impressive technology advances, dramatic cost reductions, and an increase in political support for renewable energy around the world. Not surprisingly, these industries are attracting some of the largest players in the world energy market, including BP, Royal Dutch/Shell, and General Electric (which has moved into both the wind and solar cell markets in recent years). They are even drawing other major companies—including DuPont and Honda—into the energy arena for the first time. It is against this backdrop that the alternative forms of energy are assessed and analyzed.

Solar Power

When examining the state of the industry, it is clear that for solar energy to achieve its potential, plant construction costs will have to be further reduced via technology improvements, economies of scale, and streamlined assembly techniques. Development of economic storage technologies can also lower costs significantly. The U.S. Southwest has some of the most valuable solar resources in the world, with much of this potential close to major urban areas and on land that has few if any alternative economic uses. According to the National Renewable Energy Laboratory, a solar plant covering 10 square miles of desert would produce as much power as the Hoover Dam. Desert-based power plants could well provide a large share of the nation's commercial energy. (Florida Solar Energy Center 2008)

The sun's energy could provide much of the heating and cooling for America's homes and industries. Solar water heaters, which have been used for decades, are a particularly convenient way to use the sun's energy. Simple rooftop collectors made of steel, glass, and plastic heat water, while natural gas or electricity is used for backup when the sun isn't shining. Solar systems can be used from New England to California and are more cost effective in Chicago than Miami, due to Chicago's higher energy prices.

In some climates, solar heaters can provide up to 80 percent of a home's hot water. Residential solar water heating systems initially cost between $1,500 and $3,500, compared to $150–$450 for electric and natural gas water heaters, but they typically pay for themselves in 4–8 years through fuel savings. Savings continue for the remaining 15–40 year life of the system. Newer systems with low-cost plastic polymers and highly efficient vacuum tubes are providing new options and lower costs. (SOLAR Association 2008)

The United States led the solar heating industry in the 1980s, but since then the almost complete elimination of government incentives, combined with falling natural gas prices, left the United States far behind. More than 1.5 million U.S. homes and businesses now use solar water heating, and their systems produce enough energy annually to offset the output of a nuclear power plant. Only about 8 percent of these systems are used for water and space heating; the rest heat swimming pools. Hawaii leads the nation in per capita use of solar water heating, thanks to utility rebate programs and the lack of natural gas, which have driven significant demand for residential systems.

Solar energy is being tapped for space heating in commercial and industrial buildings as well. Typically, a building's south-facing wall is covered with dark-colored perforated metal sheeting, which collects solar heat that is distributed into the building through conventional ductwork. Up to 80 percent of available solar radiation is converted to heat. Solar space heating systems are more expensive than water heating systems, but will become more competitive as conventional heating costs rise. And solar energy can be used for cooling via the oldest form of air conditioning technology— absorption cooling—with the same devices used to provide heat in the winter.

Worldwide, solar heating is booming: the global market doubled between 2000 and 2005, with the greatest increases in China and Europe. (IEA Photovoltaic Power Systems Programme 2008)Japan has led the solar PV industry for most of the past decade, despite having half the solar resource of California. Strong incentives from government policies—including gradually declining rebates, net metering, low-interest loans, and public education programs— boosted

Japan from a minor player in the early 1990s to the world's largest producer and user of solar PV within a decade. Japan's policies drove down system costs by more than 80 percent, to the point where rooftop power is now competitive with Japanese electricity prices, which are among the worlds highest. (Solar Energy Industries Association 2008)

Today, Japan remains the world's leading solar PV manufacturer, accounting for 48 percent of production in 2005, but Germany is now the leading market. High purchase prices for PV-generated electricity have been a powerful driver of German demand. Germany added an estimated 600 MW during 2005 alone—far more than cumulative U.S. installed capacity. Both Germany and Japan have reaped significant employment and economic benefits from strong policies aimed at expanding markets and driving down costs. Spain, the first country to require installation of PV in new and renovated buildings, will likely join them soon. Rapid growth in Japan and Europe has encouraged major companies—some entering the energy industry for the first time—to step up investments in solar PV. These investors include Japan's Sharp and Kyocera companies, oil giants BP and Royal Dutch/Shell, and General Electric and DuPont in the United States. ((Solar Energy Industries Association 2008)

The United States is the birthplace of the solar cell industry and, as recently as 1996, U.S. producers held 44 percent of the global solar cell market. By 2005, that figure had fallen to below 9 percent as markets boomed in other parts of the world, and U.S. producers had lost much of the market at home as well. But this trend could reverse due to new state policies driving demand. In early 2006, California state regulators approved $3.2 billion in customer rebates with the goal of installing 3,000 MW of PV on the rooftops of one million California homes, businesses, and public buildings by 2017, up from about 100 MW today. New Jersey, which offers a rebate and sales tax exemption for solar PV, has the second largest U.S. market after California. ((Solar Energy Industries Association 2008)

The International Energy Agency (IEA) estimates that PV installed on appropriate rooftops, facades, and building envelopes in the United States could meet about 55 percent of U.S. electricity demand. The Solar Energy Industries Association aims for PV to provide half of all new U.S. electricity generation by 2025; SEIA projects that by 2020, the PV industry could provide Americans with 130,000 new jobs. Beyond rooftops, solar cells can replace diesel generators for water pumping on America's farms, wastewater treatment plants, and other uses. And they can produce power on a large scale in the U.S. Southwest. According to an IEA study, very-large-scale PV systems installed on just 4 percent of the world's deserts could generate enough electricity annually to meet world power demand.

Large desert-based power plants concentrate the sun's energy to produce high-temperature heat for industrial processes or convert it into electricity that is available when demand is greatest. Resource calculations show that just seven states in the U.S. Southwest could provide more than 7 million MW of solar generating capacity—roughly 10 times the total U.S. generating capacity from all sources today. Four concentrating solar technologies are being developed. To date, parabolic trough technology provides the best performance and lowest cost of all types of solar power plants. Nine plants, totaling 354 MW, have operated reliably in California's Mojave Desert since the mid-1980s.

Dish engine and power tower systems are in earlier stages of prototype and commercial development. Natural gas and other fuels can provide supplementary heating when the sun is inadequate, allowing solar power plants to generate electricity whenever it is needed. In addition, heat-storing technologies are being developed to extend the operating times of solar power plants. Since the first 14 MW trough plant was installed in California in the early 1980s, generating costs have dropped from 45 cents/kWh (in 2005 dollars) to 9–12 cents/kWh (competitive with peak power). Costs are expected to drop to 4–7 cents/kWh by 2020. Several solar power plants are now being planned in the U.S. Southwest, spurred by state requirements that a minimum share of electricity come from solar technologies.(Rocky Mountain Institute 2008)

Marine Energy

The research into the area of marine energy technology reveals that just off America's coastlines are energy resources with the potential to contribute substantially to the U.S. economy. Oceans cover roughly 70 percent of the Earth's surface and collect and store a tremendous amount of heat from the sun as well as mechanical energy in the form of tides and waves. Seawater is about 800 times as dense as air, so even slow velocities of water contain enormous quantities of energy. Globally, wave and ocean thermal energy individually are estimated to be of the same order of magnitude as present world energy demand, while energy from tides and currents is capable of making a roughly 10 percent contribution. From the Middle Ages until the Industrial Revolution, tide mills were common sights along the coasts of Western Europe.

Today, tidal power is the most commercially advanced of the ocean energy technologies, and recent innovations in tidal power technologies avoid the environmental impacts of damming bays or estuaries. Other forms of modern marine energy conversion are still at the early stages of development, with a variety of technology types being explored. Engineers consider these technologies to be 10–20 years behind wind power, but to be coming of age rapidly. Small-scale wave and tidal current projects are now being installed around the world. Europe, Australia, and Japan are further along in development of these sources than the United States, primarily because of more extensive government support. As a result, major private investors such as Electricité de France are now involved in prototype projects. (National Hydropower Association 2008)

Recently, a few U.S. states, cities, and electric utilities have begun to fund research and commit to purchasing electricity from demonstration plants. Small projects have been proposed for the cities of New York and San Francisco and off the coasts of Massachusetts, Washington, Oregon, and Hawaii. A tidal project planned for New York's East River could eventually provide power for 8,000 homes. While ocean thermal energy and current energy are concentrated in specific areas (Hawaii for ocean thermal and Florida for current energy), most coastal states could tap their wave and tidal energy.

Ocean energy resources are generally more consistent than wind or solar energy, and offer significant potential for job creation in coastal communities where shipbuilding and commercial fishing are in decline. The Electric Power Research Institute (EPRI) estimates that U.S. near-shore wave resources alone could generate some 2.3 trillion kWh of electricity annually, or more than eight times the yearly output from U.S. hydropower dams. U.S. ocean energy developers face significant regulatory uncertainty when it comes to siting and licensing projects, which makes it difficult to obtain financing. A one-megawatt wave energy project off the coast of Washington state has faced more licensing hurdles than those confronted by most large-scale fossil fuel plants because of jurisdictional uncertainty. Marine energy is not yet economically competitive with conventional energy, but it is already attractive for islands and isolated coastal communities that are off the grid. A recent EPRI report concluded that electricity generation from wave power, for example, could be economically feasible in the near future. Ocean Power Technologies, the world's first publicly traded wave power company, claims that total costs will be 3–4 cents/kWh for 100 MW systems.(.(National Hydropower Association 2008)

Tidal power technologies harness energy from the rise and fall of the tides, using dams to trap water in a bay or estuary at high tide. When the ocean level outside the dam has fallen enough to create a sufficient pressure difference, the trapped water is returned to the sea through conventional hydroelectric turbines. Tidal power has the advantage of being fairly predictable. Such plants have been in use for decades in Canada, China, Russia, and France (where the largest system, 240 MW, is operating).

Ocean currents, such as the Gulf Stream off the U.S. East Coast, are in effect massive rivers in the world's oceans, and they represent enormous quantities of energy. Technologies that harness these energy flows look like undersea wind turbines. A handful of prototype turbines now operate in the United Kingdom and Norway, and at least two U.S. companies are developing ocean current turbines. Ocean current energy is very site-specific (in the United States, only the eastern coast of Florida has significant potential), but it has the advantage of being highly predictable. (. (National Hydropower Association 2008)

Some wave energy devices consist of a floating buoy or hinged-raft that uses pistons to pump fluid through hydraulic motors. Oscillating water column devices use the up-and-down motion of the water surface in a "capture chamber" to alternately force air out and draw it in through a pneumatic turbine. Only a few wave energy devices have been demonstrated in the ocean for more than a few months, mainly in Europe and Japan. The greatest potential is close to coastlines, often in areas with high population densities, such as the U.S. West Coast. (National Hydropower Association 2008)

Ocean Thermal Energy Conversion (OTEC) OTEC harnesses the temperature difference between sun warmed surface waters of the Tropical Ocean and deep water at near-freezing temperatures. Warm water is used to vaporize a working fluid, which expands through a turbine and is then condensed by the deep, cold-water, enabling continuous flow of vapor through the turbine to generate electricity or to split seawater into hydrogen. In the tropics, the required temperature difference is nearly constant, so OTEC can provide baseload power. Small "proof of-concept" experiments have been conducted in Hawaii and Japan, but no full-scale OTEC plants have been built. (National Hydropower Association 2008)

Wind Power

Wind resources in the United States are far more plentiful than in Europe. The U.S. wind resource is well distributed across the country, with the most abundant winds in the Great Plains, a region that has been described as a potential "Persian Gulf" of wind power. And the Department of Energy estimates that the offshore wind resource within 5–50 nautical miles of the U.S. coastline could support about 900,000 MW of wind generating capacity— an amount approaching total current U.S. electric capacity. Although much of this resource will likely remain undeveloped because of environmental concerns and competing uses, the nation's offshore wind energy potential is enormous, and much of it lies near major urban load centers.

More fully tapping that wind will require new policies to provide more-ready access to existing high-voltage transmission lines, and in the longer run, the expansion of transmission capacity to allow Great Plains wind power to reach cities in the Midwest and on the West Coast. In the meantime, sizable wind power projects are planned or being developed in states from California to New York, Texas, and Montana. The country's largest offshore wind project has been proposed off the Texas coast in the Gulf of Mexico. (European Renewable Energy Council 2008)

As with all energy technologies, there are environmental costs associated with wind power, which have generated opposition from local residents concerned about the rapid proliferation of new projects in many parts of the country. The greatest controversy has arisen from the fact that wind turbines in some locations have killed significant numbers of birds and bats. Yet housecats, vehicles, cell phone towers, buildings, and habitat loss pose far greater hazards to birds, and progress has been made in reducing bird strikes through technological changes, such as slower rotating speeds, and careful project siting.

On balance, the environmental, economic, and social benefits of wind power outweigh the costs. During 2005, wind turbines operating in the United States offset the emission of 3.5 million tons of carbon dioxide, while reducing natural gas demand for power generation by 4–5 percent. Wind farms can be permitted and built far faster than conventional power plants. And by some estimates, every 100 MW of wind capacity creates 200 construction jobs, 2–5 permanent jobs, and up to $1 million in local property tax revenue.

As new wind farms come on line, a growing number of electric utility managers are learning how to integrate an intermittent resource into their power grids. These grids are designed to routinely manage variability in demand and supply. (European Renewable Energy Council 2008)

The amount of wind power capacity that can be accommodated depends on the size of the regional grid and the flexibility of other types of generation attached to it. In both Europe and North America, electric utilities have demonstrated the ability to manage wind generation that exceeds 20 percent of total capacity. Higher shares of wind power will be possible with modest operational adjustments and better wind forecasting.

The key to achieving this potential is a strong and consistent policy framework, at both the state and federal levels. The on-again off-again tax credit for wind power and similarly intermittent state policies have undermined the stability that companies require to invest in new installations, technologies, and factories in a sustained manner. If solid and consistent policies are implemented, wind power's contribution to the U.S. electricity supply could grow rapidly. In

June 2006, the Department of Energy committed to developing an action plan with the goal of providing up to 20 percent of U.S. electricity.

Chapter Five

Summary and Conclusions

The research looks at renewable energy sources across different applications to determine relevance and applicability in each of the instances. The research also examines the viability of energy efficiency and conservation measures as a substitute for oil in the free market economy. The viability of solar wind and marine power are examined in the context of the agricultural transportation and construction sectors.

Significance

The research has significance to the field of research for three reasons 1. The information provides valuable insight to policy makers. 2. The research will add to the body of existing research and will help propel future efforts aimed at information gathering. 3. The research is germane and relevant for the issues addressed are major issues of global importance.

Limitations

The research suffers two limitations. 1. It is time bound and can only provide a glimpse into technology as we know it today. The rapid progression of renewable technology demands that the research must continue to remain relevant. 2. The research in many of these technologies is still in its infancy and theories must be formulated and continually observed in real world applications. Essentially the only limiting factors are technology, capital and time. Opportunities exist in the near term but more effort is needed in the long term.

Recommendations

The research of the future should be: 1. Directed and focused so it can carry on the examination of viable renewable energy sources. 2. Hands on with information gathered from existing installations and applications of renewable energy and conservation technologies. The recommendation is made that the short term pursuit of conservation energy efficiency technologies be emphasized during a transition to the widespread commercial application of renewable technologies. The United States Federal and State governments should redirect their energy policies in a direction which offers clear stated support of these renewable technologies along with the financial capital to push forward full scale commercialization efforts.

Conclusions

There exists major areas for improvement in the application of renewable and energy efficiency technology. The potential exist for major gains in energy savings. The present dependence on fossil fuels could easily be eliminated with a move to energy efficiency technologies and the incremental adoption of proven renewable energy technologies. Increasingly it seems clear that neither oil nor coal nor nuclear power can be counted on to meet our future energy needs, the mandate used to be imply to reduce our dependence on foreign oil but that is no longer the answer. The only reason that our dependence on foreign oil has been reduced is through the modest gains the US economy has so far achieved by implementing efficiency gains. The grandiose schemes trumpeted by politicians have historically been ineffective, rather great reliance on energy efficiency has been the key.

Developing a viable energy system while limiting fossil fuel is feasible but there is no guarantee that it will take place. Energy policy is often a morass of contradictory incentives that are not easily reformed. The government subsidizes coal mining while paying large sums to clean up the air pollution from coal burning. Yet change is possible the deregulation of oil and natural gas prices in the United States has lead to gains in energy efficiency. Government sponsored programs to help improve home weatherization has lead to massive efficiency gains in both Canada and Sweden. Programs to spur the development of photovoltaic solar technology have met with success. The lessons learned show us the stepping stones to success for the future. As a result it is now possible top put together the outlines of an energy policy which makes environmental and economic sense.

Government actions must create the conditions for a change in energy priorities. Government action will create the conditions needed for continued innovation and commercialization of new energy technologies. Government intervention is necessary to ensure that energy markets function effectively and equitably Government policies must take in to account the far reaching implications of energy policy Government policy and leadership will allow millions of consumers and private companies to take the necessary steps required to introduce the full potential of energy efficiency in to the marketplace. Government sponsored efforts will provide the necessary incentive and motivation for private companies to proceed with the large scale commercialization of energy efficiency technologies

Programs to produce millions of barrels of synthetic fuels from coal have never made it off of the drawing board. Nuclear power supplies only five percent of world energy with no significant expansion planned. The question that comes up again and again is if not coal or oil then what? The central question to be answered is what reliable alternative do we have. The key to resolving the coal oil conundrum is simple but potentially revolutionary; greatly improved energy efficiency in the short run with a shift towards renewable sources of energy (wind solar and marine) in the long run. Energy efficiency has the capability to provide economic viable substitute for 25 % of our projected energy needs. At less than the cost of new supplies indeed improved energy efficiency should be the centerpiece of all of our energy policies during the coming decades. It can be used **immediately** to limit the environmental and economic damage caused by today's energy systems and to buy time for the development of other sources of renewable energy.

Despite the limitations and the fact that renewables are not as economically viable as the immediate gains to be realized from energy efficiency the cost of renewable energy sources are declining and the technology is improving   Renewables could well make a substantial contribution to meeting energy needs if promoted in a more steadfast manner. The contribution of efficiency combined with the reductions provided by renewable will push the transition to a sustainable energy future.

America needs a fresh and innovative approach to energy policy. Today's energy system has been shaped by a century of government subsidies and regulatory support. Even today, fossil fuels receive billions of dollars of federal subsidies each year, while the health, environmental, and security costs of those fuels are paid by society at large—and are not reflected in the market price of energy. Over the past three decades, governments in the United States and abroad have experimented with a variety of policies to promote renewable energy and improve energy efficiency. Although frequent shifts in government support have hindered development, policymakers can learn much from these experiences, which will help to build a policy framework that allows renewable energy to flourish.

Across the United States and around the world, there is one clear lesson from past policy experiments: wherever renewable energy industries have emerged, government policy reforms have played a central role. The key to a bright American energy future and a new wave of economic activity and innovation is a robust partnership between government and the private sector—providing incentives to jumpstart the new energy industries while minimizing the cost to American taxpayers.

To fully utilize America's renewable energy resources, policies should be enacted that:

Establish a consistent, predictable, and long-term framework of rules and incentives. Renewable resource developers, like other capital financers, need certainty to make informed investments. Create performance-based incentives. To leverage the most energy from each dollar of public investment, incentives must be based on the amount of energy generated or saved, rather than the cost of installation. In addition, incentives should evolve over time in a predictable manner to spur investment and innovation. Incorporate external costs and benefits into energy pricing, especially the introduction of greenhouse gas cap-and-trade. The full security, economic, and environmental costs of fossil fuels, and the full benefits of renewables, are not reflected in their prices. Including the full cost associated with energy generation in pricing would encourage producers and consumers to adjust their behavior toward more sustainable practices.

The U.S. Government should reduce subsidies for fossil fuels. In recognition of the maturity of the fossil fuel industries and the public benefit of reducing fossil fuel use, subsidies to these industries should be reduced or eliminated. The government should work to enact complementary policies for energy efficiency because renewable energy and energy efficiency go hand in hand. The policies to increase energy efficiency—including stronger building codes, increased vehicle fuel economy standards, and advanced efficiency standards for appliances— should complement policies designed to expand renewable energy production.

A policy promoting increasing reliance on renewable resources also increases the need for greater regional cooperation to ensure reliability. The electricity sector is already moving in this direction, and policies to continue this regional integration should be supported. The United States should actively cooperate with and learn from the many countries that are developing renewable energy and the policies to support it. Although numerous policies meet these overarching principles, the following specific recommendations should be established immediately.

Policy Recommendations

Governments, at all appropriate levels, should:

Establish clear and long-term goals and targets for renewable energy use and energy *efficiency gains.* State and local governments should be allowed to establish more ambitious targets beyond federal requirements.

Provide long-term, low interest loans and bonds to address high upfront costs and reduce *risk.* Renewable energy sources often require higher capital expenditures and have different depreciation timeframes than traditional energy sources. Government-backed financial instruments can help bridge the gap between traditional energy financing as investors adjust to the new investment requirements of renewable energy.

Use government purchasing power together with the private sector to build large, aggregated markets for renewable energy. Policies needed in the electricity and heating sectors include: Ensure fair market access and pricing for renewable electricity. Several countries have significantly increased their share of renewable energy by the use of "feed-in" laws requiring that a fixed price be paid for each unit of renewable electricity produced for the grid. Several U.S. states have enacted or are considering similar mechanisms. Standardized interconnection procedures are also needed.

Implement siting regulations to address environmental, aesthetic, and other concerns and to reduce uncertainty for stakeholders. Like any energy project, renewable energy resources must be developed in an environmentally responsible way; currently, developers are confronted by a patchwork of regulations and guidelines that can change rapidly. The siting process should be fair and consistent.

Enact "high-performance" building codes to improve efficiency and increase the share of energy provided from decentralized renewable *sources*. California and other states and cities have demonstrated the power of rigorous building codes to increase building efficiency and promote renewable energy. Governments at all levels should commit to meeting the highest standards in all new buildings and to retrofitting older buildings during scheduled renovations.

Off the shelf technology is now available which will make immediate contributions to energy savings through utilization in energy efficiency applications. The ability of renewable energy to make a significant reliable contribution to power generation has increased to the point where a transition away from fossil fuels can begin in earnest. The costs per kilowatt hour of renewable energy is competitive with traditional fuel sources and subsidies along with a push for commercialization will provide the necessary impetus to spark a transition way from fossil fuel based sources of energy.

Definitions

A number of terms related to present energy policy will need to be explained in order to fully appreciate the depth and breadth of the analysis. Many of the terms are operationally defined through the course of research and discussion of energy efficiency and renewable energy sources.

Energy efficiency refers to technology which allows present systems to use less energy and to use present energy more effectively by wringing more energy out of our present efforts. Examples of efficiency technology would be energy efficient light bulbs or more effective insulation for use in commercial buildings.

Climate change refers to changes in the earth's atmosphere which are resulting in shifting weather patterns and changes to the earth's atmosphere

Renewable energy are fuel sources which provide electric output without relying on a finite source for power, typical examples include wind solar and tidal power.

Sustainable energy refers to sources of energy which can be used with a minimal negative environmental and economic impact this type of energy can effectively fuel energy needs without destructive environmental impacts such as acid rain or other types of pollution.

Kilowatt hours refers to power in terms of watt based usage and for the purpose of this discussion allows for comparison between alternate fuel sources.

Solar technology refers to fuel sources such as photovoltaic which are powered by the sun and provide power to the grid or individual buildings by utilizing the energy of the sun.

Hydropower refers to fuel sources which use water power as their primary means of power generation such as dams and tidal power.

Wind power refers to power sources which achieve power generation through use of wind power. An example of wind power would be a farm using windmills.

References-Energy

Alliance to Save Energy "Creating an Energy Efficient World". (2008) retrieved June 5, 2008, from http://www.ase.org

American Coalition for Ethanol. Ethanol Today Magazine (2008) retrieved May 10, 2008, from http://www.ethanol.org

American Council for Energy Efficient Economy. (2008) retrieved June 15, 2008, from http://www.aceee.org

American Council on Renewable Energy. (2008) retrieved June 15, 2008, from http://www.acore.org

American Solar Energy Society. (2008) retrieved June 7, 2008, from http://www.ases.org

American Wind Energy Association -Wind Energy Works or America. (2008) retrieved June 15, 2008, from http://www.awea.org

Appenzeller, Tim & Dimick, D.R. (2004)"Signs from Earth Global Warming Bulletins from a Warmer World" National Geographic Vol.206 no.3

Appenzeller, Tim (2007)"Big Thaw" National Geographic June Volume 211 no.6

Biomass Council. (2008) retrieved May 2, 2008, from http://www.biomasscouncil.org

Biomass Research and Development Initiative. (2008) retrieved June 4, 2008, from http://www.bioproducts-bioenergy.gov

Blatt, Harvey (2005) America's Environmental Report Card. Massachusetts: The MIT Press

Bolinger, Mark. (2007) DOE-Annual Report on US Wind Power Installation Cost and

Performance Trends. Lawrence Berkeley National Lab

Brown, Lester (2001) Eco Economy Building an Economy for the Earth. New York : W.W. Norton and Company Publishing

Center for American Progress The Progressive Pulse. (2008) retrieved May 13, 2008, from http://www.americanprogress.org

Center for Resource Solutions. (2008) retrieved May 13, 2008, from http://www.resource-solutions.org /index.htm

Clean Energy Group. (2008) retrieved April 22, 2008; from http://www.cleanegroup.org new site is pewtrusts.org

Clean Energy States Alliance. (2008) retrieved May 7, 2008, from http://www.cleanenergystates.org

Clear the Air. (2008) retrieved June 11, 2008, from http://www.cleartheair.org

Climate Solutions Practical Solutions to Global Warming-Latest News. (2008) retrieved April 22, 2008, from http://www.climatesolutions.org

Database of State Incentives for Renewable Energy DSIRE Library (2008) retrieved May 21, 2008, from   http://www.dsireusa.org

Energy Efficiency and Renewable Energy, DOE. (2008) retrieved June 5, 2008, from http://www.eere.energy.gov

Energy Future Coalition America's New Energy Future. (2008) retrieved April 22, 2008, from http://www.energyfuturecoalition.org

Environmental and Energy Study Institute EESI environmental and Energy Solutions. (2008) retrieved May 11, 2008, from http://www.eesi.org

Environmental Protection Agency. (2008) retrieved April 23, 2008, from http://www.epa.gov/

European Renewable Energy Council Documents. (2008) retrieved May 22, 2008, from
http://www.erec-renewables.org

European Union, New and Renewable Energies. (2008) retrieved July1, 2008, from
http://europa.eu.int/comm/energy/res/index_en.htm

Florida Solar Energy Center Research Section. (2008) retrieved May 15, 2008, from
http://www.fsec.ucf.edu

Geothermal Energy Association. (2008) retrieved June 8, 2008, from http://www.geo-energy.org/

Green Building Alliance Smart Solutions for the Built Environment. (2008) retrieved

June 2, 2008, from http://www.gbapgh.org

Green-e Renewable Electricity Certification Program. (2008) retrieved May 15, 2008, from
http://www.green-e.org

International Energy Agency (IEA). (2008) retrieved May 15, 2008, from http://www.iea.org

Interstate Renewable Energy Council. (2008) retrieved May 12, 2008, from
http://www.irecusa.org. /

Kunstler, J.H. (2005) The Long Emergency Surviving the End of Oil. New York: Grove Press

National Hydropower Association. (2008) retrieved June 17,2008, from http://www.hydro.org

National Renewable Energy Laboratory. (2008) retrieved June 6, 2008, from
http://www.nrel.gov Nicklen, Paul (2007). "*Vanishing Sea Ice Life at the Edge*" National
Geographic June

Volume 211 No.6

Pew Center for Climate Change. (2008) retrieved May 3, 2008, from http://www.pewclimate.org

Renewable Energy Policy Network for the 21ˢᵗ Century. (2008) retrieved June 5, 2008, from

http://www.ren21.net

Renewable Energy Policy Project REPP Library Archives. (2008) retrieved June 8, 2008, from

http://www.repp.org/

Renewable Fuels Association. (2008) retrieved May 26, 2008, from http://www.ethanolrfa.org

Rocky Mountain Institute. (2008) retrieved May 22, 2008, from http://www.rmi.org/

Solar Energy Industries Association. (2008) retrieved May 27, 2008, from http://www.seia.org

Union of Concerned Scientists. (2008) retrieved June 15, 2008, from http://www.ucsusa.org /

U.S. Green Buildings Council. (2008) retrieved May 17, 2008, from http://www.usgbc.org

Utility Wind Integration Group New and Notable. (2008) retrieved May 27, 2008, from

http://www.uwig.org

Worldwatch Institute. (2008) retrieved June 5, 2008, from http://www.worldwatch.org

Table Energy Production Costs

Energy Sources	Estimated Cost per kilowatt hour
Solar PV and Thermal	4-20 cents per kwh
Wind Power	3-10 cents per kwh
Marine Generated Power	2-10 cents per kwh
Fossil Fuel Sources Coal and Oil	4 cents per kwh

Renewable energy systems encompass a broad, diverse array of technologies, and the current status of these can vary considerably. Some technologies are already mature and economically competitive (e.g. solar and hydropower), while others need additional development to become competitive without subsidies. The table shows an overview of costs of various renewable energy technologies.